Lecture Notes in Mathematics

Edited by A. Dold and B. Eckmann

682

G. D. James

The Representation Theory of the Symmetric Groups

Springer-Verlag
Berlin Heidelberg New York 1978

Author
G. D. James
Sidney Sussex College
Cambridge CB2 3HU
Great Britain

AMS Subject Classifications (1970): 20C15, 20C20, 20C30

ISBN 3-540-08948-9 Springer-Verlag Berlin Heidelberg New York
ISBN 0-387-08948-9 Springer-Verlag New York Heidelberg Berlin

Printed in Germany

Printing and binding: Beltz Offsetdruck, Hemsbach/Bergstr.
2141/3140-543210

Preface

The representation theory of the symmetric groups was first studied by Frobenius and Schur, and then developed in a long series of papers by Young. Although a detailed study of Young's work would undoubtedly pay dividends, anyone who has attempted this will realize just how difficult it is to read his papers. The author, for one, has never undertaken this task, and so no reference will be found here to any of Young's proofs, although it is probable that some of the techniques presented are identical to his.

These notes are based on those given for a Part III course at Cambridge in 1977, and include all the basic theorems in the subject, as well as some material previously unpublished. Many of the results are easier to explain with a blackboard and chalk than with the typewritten word, since combinatorial arguments can often be best presented to a student by indicating the correct line, and leaving him to write out a complete proof if he wishes. In many places of this book we have preceded a proof by a worked example, on the principle that the reader will learn more easily by translating for himself from the particular to the general than by reading the sometimes unpleasant notation required for a full proof. However, the complete argument is always included, perhaps at the expense of supplying details which the reader might find quicker to check for himself. This is especially important when dealing with one of the central theorems, known as the Littlewood-Richardson Rule, since many who read early proofs of this Rule find it difficult to fill in the details (see [16] for a description of the problems encountered).

The approach adopted is characteristic-free, except in those places, such as the construction of the character tables of symmetric groups, where the results themselves depend upon the ground field. The reader who is not familiar with representation theory over arbitrary fields must not be deterred by this; we believe, in fact, that the ordinary representation theory is <u>easier</u> to understand by looking initially at the more general situation. Nor should he be put off by the thought that technical knowledge is required for characteristic-free representation theory, since the symmetric groups enjoy special properties which make it possible for this book to be largely self-contained. The most economical way to learn the important results without using any general theorems from representation theory is to read sections 1-5, 10-11 (noting the remarks following Example 17.17), then 15-21.

Many of the theorems rely on a certain bilinear form, and towards the end we show that this bilinear form must have been known to Young, by using it in a new construction of Young's Orthogonal Form. It is remarkable that its significance in the representation theory of the symmetric groups was only recently recognized.

I wish to express my thanks to Mrs. Robyn Bringans for her careful and patient typing of my manuscript.

G. D. James

Contents

1. BACKGROUND FROM REPRESENTATION THEORY

We shall assume that the reader is familiar with the concept of the group algebra, FG, of a finite group G over a field F, and with the most elementary properties of (unital right-)FG-modules. It is possible to prove all the important theorems in the representation theory of the symmetric group using only the following:

1.1 THEOREM <u>If M is an irreducible FG-module, then M is a composition factor of the group algebra, FG.</u>

<u>Proof</u>: Let m be a non-zero element of M. Then mFG is a non-zero submodule of M, and since M is irreducible, M = mFG. The map

$$\theta: r \to mr \quad (r \in FG)$$

is easily seen to be an FG-homomorphism from FG onto M. By the first isomorphism theorem,

$$FG/\ker\theta \cong M$$

so FG has a top composition factor isomorphic to M.

The first isomorphism theorem will appear on many occasions, because we shall work over an arbitrary field, when an FG-module can be reducible but not decomposable.

We often use certain G-invariant bilinear forms, as in the proof of a special case of Maschke's Theorem:

1.2 MASCHKE'S THEOREM <u>If G is a finite group and F is a subfield of the field of real numbers, then every FG-module is completely reducible.</u>

<u>Proof</u>: Let $e_1,\ldots,e_m$ be an F-basis for our FG-module M. Then there is a unique bilinear form ϕ on M such that

$$(e_i,e_j)\phi = 1 \text{ if } i = j, \text{ and } 0 \text{ if } i \neq j.$$

Now, a new bilinear form can be defined by

$$\langle u,v\rangle = \sum_{g\in G} (ug, vg)\,\phi \quad \text{for all } u,v \text{ in } M.$$

This form is G-invariant, in the sense that

$$\langle ug, vg\rangle = \langle u,v\rangle \quad \text{for all } g \text{ in } G.$$

Given a submodule U of M, $v \in U^{\perp}$ means, by definition, that $\langle u,v\rangle = 0$ for every u in U. But if $u \in U$, then $ug^{-1} \in U$. Thus

$$\langle u,vg\rangle = \langle ug^{-1},v\rangle = 0,$$

using the fact that our form is G-invariant. This shows that $vg \in U^{\perp}$, which is the condition required for $U^{\perp}$ to be a submodule of M.

If $u \neq 0$, then $\langle u,u\rangle \neq 0$, since F is a subfield of the field of real numbers, so $U \cap U^{\perp} = 0$. We shall prove below that $\dim U + \dim U^{\perp} = \dim M$, and therefore $U^{\perp}$ is an FG-module complementing U in M as required.

We now remind the reader of some elementary algebra involving bilinear forms.

Let M be a finite-dimensional vector space over F. The <u>dual</u> of M is the vector space of linear maps from M into F, and will be denoted by M^*. Let $e_1,\ldots,e_k$ be a basis of a subspace V, and extend to a basis $e_1,\ldots,e_m$ of M. For $1 \le j \le m$, define $\varepsilon_j \in M^*$ by $e_i\,\varepsilon_j = 1$ if $i = j$, and 0 if $i \ne j$. By considering the action on $e_1,\ldots,e_m$, we see that any element ϕ of M^* can be written uniquely as a linear combination of $\varepsilon_1,\ldots,\varepsilon_m$, thus: $\phi = (e_1\phi)\varepsilon_1 + \ldots + (e_m\phi)\varepsilon_m$. Therefore, $\varepsilon_1,\ldots,\varepsilon_m$ is a basis of M^* and

$$\dim M = \dim M^* .$$

Further, ϕ belongs to V^o, the annihilator of V, if and only if $e_1\phi = \ldots = e_k\phi = 0$. Therefore, $\varepsilon_{k+1},\ldots,\varepsilon_m$ spans V^o and

$$\dim V + \dim V^o = \dim M.$$

Suppose now that we have a symmetric bilinear form, $\langle\ ,\ \rangle$, on M which is <u>non-singular</u> (That is, for every non-zero m in M there is an m' in M with $\langle m,m'\rangle \ne 0$). Define

$$\theta\colon M \to M^* \text{ by } m \to \psi_m \text{ where}$$

$$\psi_m\colon x \to \langle m,x\rangle \qquad (x \in M).$$

We see that $\psi_m \in M^*$, since $\langle\ ,\ \rangle$ is linear in the second place, and θ is a linear transformation, since $\langle\ ,\ \rangle$ is linear in the first place. Now, $\ker\theta = \{m \in M \mid \text{for all } x \in M,\ \langle m,x\rangle = 0\} = 0$, since the bilinear form is non-singular. But $\dim M = \dim M^*$, so θ is an isomorphism between M and M^*. Under this identification, $V^\perp$ corresponds to V^o. Thus, for every subspace V,

<u>1.3 $\dim V + \dim V^\perp = \dim M$</u>

Since $V \subseteq V^{\perp\perp}$, this equation between dimensions gives

$$V^{\perp\perp} = V.$$

More generally, given subspaces $0 \subseteq U \subseteq V \subseteq M$, we have $V^\perp \subseteq U^\perp$, and we may define

$$\theta\colon V \to \text{dual of } U^\perp/V^\perp \text{ by } v \to \psi_v, \text{ where}$$

$$\psi_v\colon\ x + V^\perp \to \langle v,x\rangle \qquad (x \in U^\perp).$$

If $x + V^\perp = x' + V^\perp$, then $x - x' \in V^\perp$, and $\langle v,x\rangle - \langle v,x'\rangle = \langle v,x-x'\rangle = 0$. This shows that ψ_v is well-defined. In the same way as before, ψ_v and θ are linear, but now

$$\ker\theta = \{v \in V \mid \text{for all } x \in U^\perp,\ \langle v,x\rangle = 0\} = V \cap U^{\perp\perp} .$$

Since $U^{\perp\perp} = U \subseteq V$, $\ker\theta = U$. We therefore have a monomorphism from $V/\ker\theta = V/U$ into the dual of $U^\perp/V^\perp$. Again, dimensions give:

<u>1.4 When $0 \subseteq U \subseteq V \subseteq M$, $V/U \cong$ dual of $U^\perp/V^\perp$. In particular, $V \cong$ dual of $M/V^\perp$</u>.

If M is an FG-module for the group G, we can turn the dual space

M^* into an FG-module by letting

$$m(\psi g) = (m g^{-1})\psi \qquad (m \in M,\ \psi \in M^*,\ g \in G).$$

Notice that the inverse of g appears to ensure that $\psi(gh) = (\psi g)h$. This means that the module M^* (which we shall call <u>the dual of M</u>) is not in general FG-isomorphic to M. Indeed, if T(g) is the matrix representing g with respect to the basis $e_1,\ldots,e_m$ of M, then $T'(g^{-1})$ is the matrix representing g with respect to the dual basis $\varepsilon_1,\ldots,\varepsilon_m$ of M^*. This means that the character of M^* is the complex conjugate of the character of M when we are working over the complex numbers.

Now assume that the bilinear form < , > is G invariant. If U and V are FG-submodules of M, then the isomorphisms in 1.4 are FG-isomorphisms. To verify this, we must show that $\theta: v \to \psi_v$ is a G-homomorphism. But $(x + V^\perp)\psi_{vg} = \langle x,vg\rangle = \langle xg^{-1},v\rangle = (xg^{-1} + V^\perp)\psi_v = (x + V^\perp)g^{-1}\psi_v = (x + V^\perp)(\psi_v g)$, and $\psi_{vg} = \psi_v g$, as required.

For every pair of subspaces U and V of M, $(U + V)^\perp = U^\perp \cap V^\perp$, as can easily be deduced from the definitions. Replacing U and V by $U^\perp$ and $V^\perp$, we also find that $U^\perp + V^\perp = (U \cap V)^\perp$.

Throughout this book, the next picture will be useful:

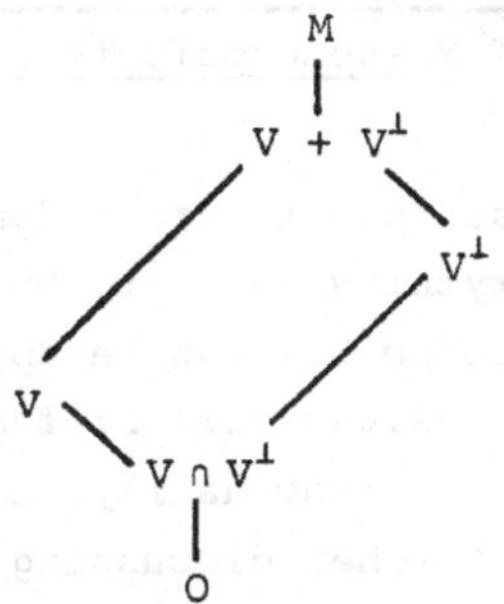

The second isomorphism theorem gives $V/(V \cap V^\perp) \cong (V + V^\perp)/V^\perp$. But $(V + V^\perp)/V^\perp \cong$ dual of $V/(V + V^\perp)^\perp$, by 1.4 = dual of $V/(V \cap V^\perp)$, so

<u>1.5</u> <u>For every FG-submodule V of M, $V/(V \cap V^\perp)$ is a self-dual FG-module</u>.

Every irreducible representation of the symmetric group will turn up in this fashion.

It is very important to notice that $V \cap V^\perp$ can be non-zero for a submodule V of M. How can we compute the dimension of $V/(V \cap V^\perp)$, given a basis of V? The answer is simple in theory, but will require a lot of calculation if V has large dimension. The <u>Gram matrix</u>, A, is defined with respect to a basis $e_1,\ldots,e_k$ of V by letting the (i,j)th entry of A be $\langle e_i,e_j\rangle$.

<u>1.6</u> <u>THEOREM</u> <u>The dimension of $V/(V \cap V^\perp)$ equals to the rank of the</u>

Gram matrix with respect to a given basis of V.

Proof: As usual, map V → dual of V by

$$\Theta: v \to \psi_v \quad \text{where} \quad u\psi_v = \langle v,u\rangle \qquad (u \in V)$$

Let $e_1,\ldots,e_k$ be the given basis of V, and $\varepsilon_1,\ldots, \varepsilon_k$ be the dual basis of V^*. Since $e_j\psi_{e_i} = \langle e_i,e_j\rangle$, we have

$$\psi_{e_i} = \langle e_i,e_1\rangle\, \varepsilon_1+\ldots+\langle e_i,e_k\rangle\, \varepsilon_k \ .$$

Thus the Gram matrix for the basis $e_1,\ldots,e_k$ coincides with the matrix of Θ taken with respect to the bases $e_1,\ldots,e_k$ of V and $\varepsilon_1,\ldots$ ε_k of V^*. But, visibly, $\ker \Theta = V \cap V^{\perp}$, so $\dim V/(V \cap V^{\perp}) = \dim \operatorname{Im} \Theta =$ the rank of the Gram matrix.

The only results from general representation theory which we shall use without proof are those telling us how many inequivalent ordinary and p-modular irreducible representations a finite group possesses, and the following well-known result about representations of a finite group over $\mathbb{C}$, the field of complex numbers (cf. Curtis and Reiner [2] 43.18 and Exercise 43.6).

1.7 Let S be an irreducible $\mathbb{C}G$-module, and M be any $\mathbb{C}G$-module. Then the number of composition factors of M isomorphic to S equals $\dim \operatorname{Hom}_{\mathbb{C}G}(S,M)$.

In fact, it turns out that these results are redundant in our approach, and Theorem 1.1 gives everything we want, but it would be foolish to postpone proofs until Theorem 1.1 can be applied.

Readers interested in character values will be familiar with the Frobenius Reciprocity Theorem and the orthogonality relations for characters, so we assume these results when discussing characters.

2. THE SYMMETRIC GROUP

The proofs of the results stated in this section can be found in any elementary book on group theory.

A function from $\{1,2,\dots,n\}$ onto itself is called a <u>permutation</u> of n numbers, and the set of all permutations of n numbers, together with the usual composition of functions, is the <u>symmetric group</u> of degree n, which will be denoted by $\mathfrak{S}_n$. Note that $\mathfrak{S}_n$ is defined for $n \geq 0$, and $\mathfrak{S}_n$ has n! elements (where 0! = 1). If X is a subset of $\{1,2,\dots,n\}$, we shall write $\mathfrak{S}_X$ for the subgroup of $\mathfrak{S}_n$ which fixes every number outside X.

It is common practice to write a permutation π as follows:

$$\pi = \begin{pmatrix} 1 & 2 & 3 & \dots & n \\ 1\pi & 2\pi & 3\pi & & n\pi \end{pmatrix}$$

By considering the orbits of the group generated by π, it is simple to see that π can be written as a product of disjoint cycles, as in the example :

$$\begin{pmatrix} 1 & 2 & 3 & 4 & 5 & 6 & 7 & 8 & 9 \\ 3 & 5 & 1 & 9 & 6 & 8 & 7 & 2 & 4 \end{pmatrix} = (2\ 5\ 6\ 8)(1\ 3)(4\ 9)(7)$$

We usually suppress the 1-cycles when writing a permutation. For example, if π interchanges the different numbers a,b and leaves the other numbers fixed, then π is called a <u>transposition</u> and is written as $\pi = (a\ b)$.

All our maps will be written <u>on the right</u>; in this way, we have $(1\ 2)(2\ 3) = (1\ 3\ 2)$. This point must be noted carefully, as some mathematicians would interpret the product as $(1\ 2\ 3)$.

Since $(i_1\ i_2 \dots i_k) = (i_1\ i_2)(i_1\ i_3)\dots(i_1\ i_k)$, any cycle, and hence any permutation, can be written as a product of transpositions. Better still,

<u>2.1</u> <u>The transpositions $(x-1,x)$ with $1 < x \leq n$ generate $\mathfrak{S}_n$.</u>

This is because, when $a < b$, we can conjugate $(b-1,b)$ by $(b-2,b-1)$ $(b-3,b-2)\dots(a,a+1)$ to obtain $(a\ b)$.

If $\pi = \sigma_1\sigma_2\dots\sigma_j = \tau_1\tau_2\dots\tau_k$ are two ways of writing π as a product of transpositions, then it can be proved that $j - k$ is even. Hence there is a well-defined function

$$\mathrm{sgn}\colon \mathfrak{S}_n \to \{\pm 1\}$$

such that $\mathrm{sgn}\ \pi = (-1)^j$ if π is a product of j transpositions.

2.2 DEFINITION $\lambda = (\lambda_1,\lambda_2,\lambda_3,\dots)$ is a <u>partition</u> of n if $\lambda_1,\lambda_2,\lambda_3,\dots$ are non-negative integers, with $\lambda_1 \geq \lambda_2 \geq \lambda_3 \geq \dots$ and $\sum_{i=1}^{\infty} \lambda_i = n$.

The permutation π is said to have <u>cycle-type</u> λ if the orbits of the group generated by π have lengths $\lambda_1 \geq \lambda_2 \geq \dots$ Thus, (2 5 6 8)(1 3)(4 9)(7) has cycle-type (4,2,2,1,0,0,...). Abbreviations such as the following will usually be adopted:

$$(4,2,2,1,0,0,\dots) = (4,2,2,1) = (4,2^2,1).$$

That is, we often suppress the zeros at the end of λ, and indicate repeated parts by an index.

Since two permutations are conjugate in $\mathfrak{S}_n$ if and only if the permutations have the same cycle type,

2.3 <u>The number of conjugacy classes of $\mathfrak{S}_n$ equals the number of partitions of n.</u>

Now, for any finite group G, the number of inequivalent irreducible $\mathbb{C}$G-modules is equal to the number of conjugacy classes of G, so

2.4 <u>The number of inequivalent ordinary irreducible representations of $\mathfrak{S}_n$ equals the number of partitions of n.</u>

We should therefore aim to construct a representation of $\mathfrak{S}_n$ for each partition of n. Let us look first at an easy example:

2.5 EXAMPLE There is a natural representation which arises directly from the fact that $\mathfrak{S}_n$ permutes the numbers $1,2,\dots,n$; take a vector space over F of dimension n, with basis elements called $\bar{1},\bar{2},\dots,\bar{n}$, and let $\mathfrak{S}_n$ act on the space by $\bar{i}\,\pi = \overline{i\pi}$ ($\pi \in \mathfrak{S}_n$). We shall denote this representation by $M^{(n-1,1)}$.

We can easily spot a submodule of $M^{(n-1,1)}$; the space U spanned by $\bar{1} + \bar{2} + \dots + \bar{n}$ is a submodule on which $\mathfrak{S}_n$ acts trivially. It is not hard to find another submodule, but suppose we wish to eliminate guesswork. If $F = \mathbb{Q}$, the field of rational numbers, the proof of Maschke's Theorem suggests we construct an $\mathfrak{S}_n$-invariant inner product on $M^{(n-1,1)}$ and then $U^\perp$ will be an invariant complement to U.

$$\langle \bar{i},\bar{j}\rangle = 1 \text{ if } i = j \text{ and } 0 \text{ if } i \neq j \qquad (*)$$

defines an $\mathfrak{S}_n$-invariant inner product on $M^{(n-1,1)}$. Then

$$U^\perp = \{\textstyle\sum a_i\,\bar{i} \mid a_i \in \mathbb{Q} \quad a_1 + \dots + a_n = 0\}.$$

Let $S^{(n-1,1)} = (\bar{2} - \bar{1})F\mathfrak{S}_n$. Then certainly $S^{(n-1,1)}$ is a submodule of $U^\perp$, and it is easy to see that we have equality. Thus

$$M^{(n-1,1)} = S^{(n-1,1)} \oplus U \quad \text{when } F = \mathbb{Q}.$$

Notice though, that (*) gives an $\mathfrak{S}_n$-invariant bilinear form on $M^{(n-1,1)}$ <u>whatever the field</u>. $S^{(n-1,1)}$ is always a submodule, too (It is a complement to U if and only if char $F \nmid n$.) $S^{(n-1,1)}$ is a <u>Specht module.</u>

Are there any other easy ways of constructing representation modules for $\mathfrak{S}_n$? Consider the vector space $M^{(n-2,2)}$, over F spanned by unordered pairs $\overline{ij}$ $(i \neq j)$. $M^{(n-2,2)}$ has dimension $\binom{n}{2}$, and becomes an $F\mathfrak{S}_n$-module if we define $\overline{ij}\,\pi = \overline{i\pi, j\pi}$. This space should not be difficult to handle, but it is not irreducible, since $\sum\{\overline{ij} \mid 1 \le i < j \le n\}$ is a trivial submodule. We do not go into details for the moment, but simply observe that $M^{(n-2,2)}$ supplies more scope for investigation.

More generally, we can work with the vector space $M^{(n-m,m)}$ spanned by unordered m-tuples $\overline{i_1 \ldots i_m}$ (where $i_j \neq i_k$ unless $j = k$). Since this space is isomorphic to that spanned by unordered (n-m)-tuples, there is no loss in assuming that $n-m \ge m$. This means that for every partition of n with two non-zero parts we have a corresponding (reducible) $F\mathfrak{S}_n$-module at our disposal.

Flushed with this success, we should go on and see what else we can do. Let $M^{(n-2,1^2)}$ be the space spanned by ordered pairs, which we shall denote by $\frac{\overline{i}}{\underline{j}}$ $(i \neq j)$. The $\mathfrak{S}_n$ action is $\frac{\overline{i}}{\underline{j}}\,\pi = \frac{\overline{i\pi}}{\underline{j\pi}}$. Let $M^{(n-3,2,1)}$ be the space spanned by vectors consisting of an unordered 2-tuple $\overline{ij}$ followed by a 1-tuple k, where no two of i,j and k are equal. These vectors may be denoted by $\frac{\overline{ij}}{\underline{k}}$, but it seems that we should change our notation and have

$$\frac{\overline{i_1 \cdots\cdots i_{n-3}}}{\frac{\overline{i_{n-2}\ i_{n-1}}}{\underline{i_n}}}$$

as a basis vector of $M^{(n-3,2,1)}$ in place of

$$\frac{\overline{i_{n-2}\ i_{n-1}}}{\underline{i_n}}\ .$$

By now, it should be clear how to construct an $F\mathfrak{S}_n$-module M^λ for each partition λ of n. The notation we need to do this formally is introduced in the next section. M^λ is reducible (unless $\lambda = (n)$), but contains a Specht module S^λ, which it turns out, is irreducible if char $F = 0$.

3. DIAGRAMS, TABLEAUX AND TABLOIDS

<u>3.1</u> DEFINITIONS. If λ is a partition of n, then the <u>diagram</u> $[\lambda]$ is $\{(i,j) \mid i,j \in \mathbb{Z} \quad 1 \le i \quad 1 \le j \le \lambda_i\}$ (Here, $\mathbb{Z}$ is the set of integers). If $(i,j) \in [\lambda]$, then (i,j) is called a <u>node</u> of $[\lambda]$. The kth <u>row</u> (respectively, <u>column</u>) of a diagram consists of those nodes whose first (respectively, second) coordinate is k.

We shall draw diagrams as in the following example:

```
                          x x x x
λ = (4,2²,1)      [λ] =   x x
                          x x
                          x
```

There is no universal convention about which way round diagrams should be shown. Some mathematicians work with their first coordinate axis to the right and the second one upwards! It is customary to drop the inner brackets when giving examples of diagrams, so we write $[4,2^2,1]$, not $[(4,2^2,1)]$.

The set of partitions of n is partially ordered by

3.2 DEFINITION. If λ and μ are partitions of n, we say that λ <u>dominates</u> μ, and write $\lambda \trianglerighteq \mu$, provided that

$$\text{for all } j, \quad \sum_{i=1}^{j} \lambda_i \ge \sum_{i=1}^{j} \mu_i$$

If $\lambda \trianglerighteq \mu$ and $\lambda \ne \mu$, we write $\lambda \triangleright \mu$.

3.3 EXAMPLE. The dominance relation on the set of partitions of 6 is shown by the tree:

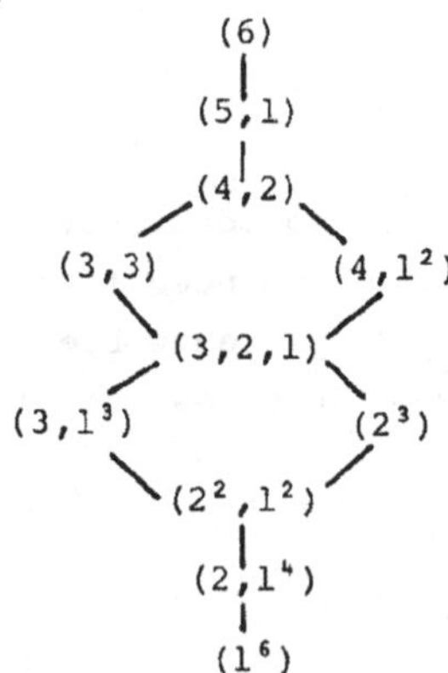

The dominance order is certainly the "correct" order to use for partitions, but it is sometimes useful to have a total order, >, on the set of partitions. The one we use is given by

3.4 DEFINITION If λ and μ are partitions of n, write $\lambda > \mu$ if and only if the least j for which $\lambda_j \ne \mu_j$ satisfies $\lambda_j > \mu_j$. (Note that some authors write this relation as $\lambda < \mu$). This is called the <u>dictionary order</u> on partitions.

It is simple to verify that the total order $>$ contains the partial order $\trianglerighteq$, in the sense that $\lambda \trianglerighteq \mu$ implies $\lambda > \mu$. But the reverse implication is false since

$(6)>(5,1)>(4,2)>(4,1^2)>(3^2)>(3,2,1)>(3,1^3)>(2^3)>(2^2,1^2)>(2,1^4)>(1^6)$.

3.5 DEFINITION If $[\lambda]$ is a diagram, the <u>conjugate</u> diagram $[\lambda']$ is obtained by interchanging the rows and columns in $[\lambda]$. λ' is the partition of n conjugate to λ.

The only use of the total order $>$ is to specify, say, the order in which to take the rows of the character table of $\mathfrak{S}_n$. Since there may be more than one self-conjugate partition of n (e.g. $(4,2,1^2)$ and $(3^2,2)$ are both self-conjugate partitions of 8), there is no "symmetrical" way of totally ordering partitions, so that the order is reversed by taking conjugates. It is interesting to see, though, that

$$\lambda \trianglerighteq \mu \text{ if and only if } \mu' \trianglerighteq \lambda'.$$

The next thing to define is a λ-tableau. This can be defined as a bijection from $[\lambda]$ to $\{1,2,\ldots,n\}$, but we prefer the less formal

3.6 DEFINITION A <u>λ-tableau</u> is one of the n! arrays of integers obtained by replacing each node in $[\lambda]$ by one of the integers $1,2,\ldots,n$, allowing no repeats.

For example,

```
1 2 4 5     and     4 5 7 3     are (4,3,1)-tableaux.
3 6 7               2 1 8
8                   6
```

$\mathfrak{S}_n$ acts on the set of λ-tableaux in the natural way; thus the permutation (1 4 7 8 6)(2 5 3) sends the first of the tableaux above to the second. (Of course, the definition of a tableau as a function wins here. Given a tableau t and a permutation π, the compositions of the functions t and π gives the new tableau $t\pi$).

Every approach to the representation theory of $\mathfrak{S}_n$ depends upon a form of the next result, which relates the dominance order on partitions to a property of tableaux.

3.7 THE BASIC COMBINATORIAL LEMMA <u>Let λ and μ be partitions of n, and suppose that t_1 is a λ-tableau and t_2 is a μ-tableau. Suppose that for every i the numbers from the ith row of t_2 belong to different columns of t_1. Then $\lambda \trianglerighteq \mu$.</u>

<u>Proof</u>: Imagine that we can place the μ_1 numbers from the first row of t_2 in $[\lambda]$ such that no two numbers are in the same column. Then $[\lambda]$ must have at least μ_1 columns; that is $\lambda_1 \geq \mu_1$. Next insert the μ_2

numbers from the second row of t_2 in different columns. To have space to so this, we require $\lambda_1 + \lambda_2 \geq \mu_1 + \mu_2$. Continuing in this way, we have $\lambda \trianglerighteq \mu$.

3.8 DEFINITIONS If t is a tableau, its <u>row-stabilizer</u>, R_t, is the subgroup of $\mathfrak{S}_n$ keeping the rows of t fixed setwise.

i.e. $R_t = \{\pi \in \mathfrak{S}_n \mid \text{for all } i,\ i \text{ and } i\pi \text{ belong to the same row of } t\}$

The <u>column stabilizer</u> C_t, of t is defined similarly.

For example, when $t = \begin{matrix} 1 & 2 & 4 & 5 \\ 3 & 6 & 7 & \\ 8 & & & \end{matrix}$, $R_t = \mathfrak{S}_{\{1\ 2\ 4\ 5\}} \times \mathfrak{S}_{\{3\ 6\ 7\}} \times \mathfrak{S}_{\{8\}}$

and $|R_t| = 4!\ 3!\ 1!$

Note that $R_{t\pi} = \pi^{-1}R_t\pi$ and $C_{t\pi} = \pi^{-1}C_t\pi$.

3.9 DEFINITION Define an equivalence relation on the set of λ-tableaux by $t_1 \sim t_2$ if and only if $t_1\pi = t_2$ for some $\pi \in R_{t_1}$. The <u>tabloid</u> $\{t\}$ containing t is the equivalence class of t under this equivalence relation.

It is best to regard a tabloid as a "tableau with unordered row entries". In examples, we shall denote $\{t\}$ by drawing lines between the rows of t. Thus

3 4 5	2 4 5	1 4 5	2 3 5	1 3 5	1 2 5	2 3 4	1 3 4	1 2 4	1 2 3
1 2	1 3	2 3	1 4	2 4	3 4	1 5	2 5	3 5	4 5

are the different (3,2)-tabloids, and $\begin{array}{c}\hline 1\ 3\ 2 \\ \hline 5\ 4 \\ \hline\end{array} = \begin{array}{c}\hline 1\ 2\ 3 \\ \hline 4\ 5 \\ \hline\end{array}$.

$\mathfrak{S}_n$ acts on the set of λ-tabloids by $\{t\}\pi = \{t\pi\}$. This action is well-defined, since $\{t_1\} = \{t_2\}$ implies $t_2 = t_1\sigma$ for some σ in R_{t_1}. Then $\pi^{-1}\sigma\pi \in \pi^{-1}R_{t_1}\pi = R_{t_1\pi}$, so $\{t_1\pi\} = \{t_1\sigma\pi\} = \{t_2\pi\}$.

We totally order the λ-tabloids by

3.10 DEFINITION $\{t_1\} < \{t_2\}$ if and only if for some i

(i) When $j > i$, j is in the same row of $\{t_1\}$ and $\{t_2\}$

(ii) i is in a higher row of $\{t_1\}$ than $\{t_2\}$.

We have written the (3,2)-tabloids in this order, above. There are many other sensible orderings of λ-tabloids, but the chosen method is sufficient for most of our purposes. As with the dominance order on partitions, the best tabloid ordering is a partial one:

3.11 DEFINITION Given any tableau t, let $m_{ir}(t)$ denote the number of entries less than or equal to i in the first r rows of t. Then write

$\{t_1\} \trianglelefteq \{t_2\}$ if and only if for all i and r $m_{ir}(t_1) \leq m_{ir}(t_2)$.

This orders the tabloids of all shapes and sizes, but we shall compare only tabloids associated with the same partition.

By considering the largest i, then the largest r, such that $m_{ir}(t_1) < m_{ir}(t_2)$, it follows that

<u>3.12</u> <u>For λ-tabloids $\{t_1\}$ and $\{t_2\}$, $\{t_1\} \triangleleft \{t_2\}$ implies $\{t_1\} < \{t_2\}$.</u>

3.13 EXAMPLES (i) If $t_1 = \begin{matrix} 1 & 3 & 6 \\ 2 & 5 & 7 \\ 4 & & \end{matrix}$ and $t_2 = \begin{matrix} 1 & 2 & 4 \\ 3 & 5 & 6 \\ 7 & & \end{matrix}$

then the first 7 rows and 3 columns of the matrices $(m_{ir}(t_1))$ and $(m_{ir}(t_2))$ are

$$(m_{ir}t_1)) = \begin{matrix} 1 & 1 & 1 \\ 1 & 2 & 2 \\ 2 & 3 & 3 \\ 2 & 3 & 4 \\ 2 & 4 & 5 \\ 3 & 5 & 6 \\ 3 & 6 & 7 \end{matrix} \qquad (m_{ir}(t_2)) = \begin{matrix} 1 & 1 & 1 \\ 2 & 2 & 2 \\ 2 & 3 & 3 \\ 3 & 4 & 4 \\ 3 & 5 & 5 \\ 3 & 6 & 6 \\ 3 & 6 & 7 \end{matrix}$$

Therefore, $\{t_1\} \triangleleft \{t_2\}$.

(ii) The tree below shows the $\triangleleft$ relation on the (3,2)-tabloids:

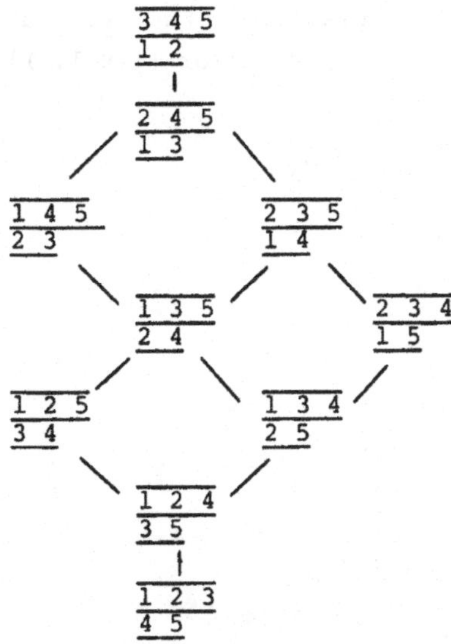

Suppose that w < x and w is in the ath row and x is in the bth row of t. Then the definition of $m_{ir}(t)$ gives

<u>3.14</u> $$m_{ir}(t(wx)) - m_{ir}(t) = \begin{cases} 1 \text{ if } b \le r < a \text{ and } w \le i < x \\ -1 \text{ if } a \le r < b \text{ and } w \le i < x \end{cases}$$

$\quad 0$ otherwise.

Therefore

<u>3.15</u> $\{t\} \lhd \{t(w\,x)\}$ if $w < x$ and w is lower than x in t.

When we prove Young's Othogonal Form, we shall need to know that the tabloids $\{t\}$ and $\{t(x-1,x)\}$ are immediately adjacent in the $\lhd$ order (or are the same tabloid):

3.16 LEMMA If $x-1$ is lower than x in t, and t is a λ-tableau, then there is no λ-tableau t_1 with $\{t\} \lhd \{t_1\} \lhd \{t(x-1,x)\}$.

Proof: First note that for any tableau t^* with i^* in the r^*th row, $m_{i^*r}(t^*) - m_{i^*-1,r}(t^*)$ = the number of numbers equal to i^* in the first r rows of t^* = $\begin{cases} 0 & \text{if } r < r^* \\ 1 & \text{if } r \geq r^* \end{cases}$

Now suppose that $x-1$ is lower than x in t, and $\{t\} \unlhd \{t_1\} \unlhd \{t(x-1,x)\}$. By 3.14,

$$m_{ir}(t) = m_{ir}(t(x-1,x)) \quad \text{if } i \neq x-1.$$

Therefore $\quad m_{ir}(t_1) = m_{ir}(t) \quad \text{if } i \neq x-1$

and $\quad m_{ir}(t) - m_{i-1,r}(t) = m_{ir}(t_1) - m_{i-1,r}(t_1)$ if $i \neq x-1$ or x.

By the first paragraph of the proof, all the numbers except $x-1$ and x appear in the same place in t and t_1. But t and t_1 are both λ-tableaux. Therefore, $\{t_1\} = \{t\}$ or $\{t(x-1,x)\}$ as required.

4. SPECHT MODULES

With each partition μ of n, we associate a Young subgroup $\mathfrak{S}_\mu$ of $\mathfrak{S}_n$ by taking

$$\mathfrak{S}_\mu = \mathfrak{S}_{\{1,2,\ldots,\mu_1\}} \times \mathfrak{S}_{\{\mu_1+1,\ldots,\mu_1+\mu_2\}} \times \mathfrak{S}_{\{\mu_1+\mu_2+1,\ldots,\mu_1+\mu_2+\mu_3\}} \times$$

The study of representations of $\mathfrak{S}_n$ starts with the permutation module M^μ of $\mathfrak{S}_n$ on $\mathfrak{S}_\mu$. The Specht module S^μ is a submodule of M^μ, and when the base field is $\mathbb{Q}$ (the field of rational numbers), the different Specht modules, as μ varies over partitions of n, give all the ordinary irreducible representations of $\mathfrak{S}_n$.

4.1 DEFINITION Let F be an arbitrary field, and let M^μ be the vector space over F whose basis elements are the various μ-tabloids.

The action of $\mathfrak{S}_n$ on tabloids has already been defined, by $\{t\}\pi = \{t\pi\}$ ($\pi \in \mathfrak{S}_n$). Extending this action to be linear on M^μ turns M^μ into an $F\mathfrak{S}_n$-module, and because $\mathfrak{S}_n$ is transitive on tabloids, with $\mathfrak{S}_\mu$ stabilizing one tabloid,

4.2 <u>M^μ is the permutation module of $\mathfrak{S}_n$ on the subgroup $\mathfrak{S}_\mu$. M^μ is a cyclic $F\mathfrak{S}_n$-module, generated by any one tabloid, and dim $M^\mu = n!/(\mu_1!\mu_2!\ldots.)$.</u>

4.3 DEFINITIONS Suppose that t is a tableau. Then the <u>signed column sum</u>, κ_t, is the element of the group algebra $F\mathfrak{S}_n$ obtained by summing the elements in the column stabilizer of t, attaching the signature to each permutation. In short,

$$\kappa_t = \sum_{\pi \in C_t} (\operatorname{sgn} \pi)\pi \ .$$

The <u>polytabloid</u>, e_t, associated with the tableau t is given by

$$e_t = \{t\}\kappa_t$$

The <u>Specht module</u> S^μ for the partition μ is the submodule of M^μ spanned by polytabloids.

A polytabloid, it must be noted, depends on the tableau t, not just the tabloid $\{t\}$. All the tabloids involved in e_t have coefficient ± 1 (If $v \in M^\mu$, then v is a linear combination of tabloids; we say that the tabloid $\{t\}$ is <u>involved</u> in v if its coefficient is non-zero.)

4.4 EXAMPLE If $t = \begin{matrix} 2 & 5 & 1 \\ 3 & 4 & \end{matrix}$ then $\kappa_t = (1-(2\ 3))(1-(4\ 5))$.

(We always denote the identity permutation by 1). Also

$$e_t = \overline{\underline{\begin{matrix} 2 & 5 & 1 \\ 3 & 4 & \end{matrix}}} - \overline{\underline{\begin{matrix} 3 & 5 & 1 \\ 2 & 4 & \end{matrix}}} - \overline{\underline{\begin{matrix} 2 & 4 & 1 \\ 3 & 5 & \end{matrix}}} + \overline{\underline{\begin{matrix} 3 & 4 & 1 \\ 2 & 5 & \end{matrix}}}$$

The practical way of writing down e_t, given t, is to permute the numbers in the columns of t in all possible ways, attaching the signature of the relevant permutation to each tableau obtained that way, and then draw lines between the rows of each tableau.

Since $\kappa_t\pi = \pi\kappa_{t\pi}$, we have $e_t\pi = e_{t\pi}$, so

4.5 S^μ is a cyclic module, generated by any one polytabloid.

It we wish to draw attention to the ground field F, we shall write M^μ_F and S^μ_F. Many results for Specht modules work over an integral domain, and it is only in Theorem 4.8 and Lemma 11.3 that we must have a field. When F is unspecified, then the ground field is arbitrary. Since M^μ is a permutation module, it is hardly surprising that most of its properties (for instance, its dimension) are independent of the base field. What is more remarkable is that many results for the Specht module are also independent of the field. Two special cases are immediate. When $\mu = (n)$, $S^\mu = M^\mu$ = the trivial $F\mathfrak{S}_n$-module. When $\mu = (1^n)$, M^μ is isomorphic to the regular representation of $\mathfrak{S}_n$, and S^μ is the alternating representation (i.e. $\pi \to \text{sgn}\pi$).

We now use the basic combinatorial Lemma 3.7 to prove

4.6 LEMMA Let λ and μ be partitions of n. Suppose that t is a λ-tableau and t^* is a μ-tableau, and that $\{t^*\}\kappa_t \neq 0$. Then $\lambda \trianglerighteq \mu$, and if $\lambda = \mu$ then $\{t^*\}\kappa_t = \pm\{t\}\kappa_t$ $(= \pm e_t)$.

Proof: Let a and b be two numbers in the same row of t^*. Then

$$\{t^*\}\ (1-(a\ b)) = \{t^*\} - \{t^*(a\ b)\} = 0.$$

a and b cannot be in the same column of t, otherwise we could select signed coset representatives $\sigma_1,\ldots,\sigma_k$ for the subgroup of the column stabilizer of t consisting on 1 and (a,b) and obtain

$$\kappa_t = (1-(a\ b))(\sigma_1 + \ldots+\sigma_k).$$

It would then follow that $\{t^*\}\kappa_t = 0$, contradicting our hypothesis.

We have now proved that for every i, the numbers in the ith row of t^* belong to different columns of t, and Lemma 3.7 gives $\lambda \trianglerighteq \mu$. Also, if $\lambda = \mu$, then $\{t^*\}$ is one of the tabloids involved in $\{t\}\kappa_t$, by construction. Thus, in this case, $\{t^*\} = \{t\}\pi$ for some permutation π in C_t, and $\{t^*\}\kappa_t = \{t\}\pi\ \kappa_t = \pm\{t\}\kappa_t$.

4.7 COROLLARY It u is an element of M^μ and t is a μ-tableau, then $u\kappa_t$ is a multiple of e_t.

Proof: u is a linear combination of μ-tabloids $\{t^*\}$ and $\{t^*\}\kappa_t$ is a multiple of e_t, by the Lemma.

Now let $< \ ,\ >$ be the unique bilinear form on M^μ for which $< \{t_1\},\{t_2\} > = 1$ if $\{t_1\} = \{t_2\}$, 0 if $\{t_1\} \neq \{t_2\}$.

Clearly, this is a symmetric, $\mathfrak{S}_n$-invariant, non-singular bilinear form on M^μ, whatever the field. If the field is $\mathbb{Q}$, then the form is an inner product (cf. Example 2.5).

We shall often use the following trick:

For $u, v \in M^\mu$,
$$\begin{aligned} \langle u\kappa_t, v\rangle &= \sum_{\pi\in C_t} \langle (\operatorname{sgn}\pi)u\pi, v\rangle \\ &= \sum_{\pi\in C_t} \langle u, (\operatorname{sgn}\pi)v\pi^{-1}\rangle \\ &\quad \text{(since the form is } \mathfrak{S}_n\text{-invariant.)} \\ &= \sum_{\pi\in C_t} \langle u, (\operatorname{sgn}\pi)v\pi\rangle \\ &= \langle u, v\kappa_t\rangle \ . \end{aligned}$$

The crucial result using our bilinear form is

4.8 THE SUBMODULE THEOREM (James [7]). <u>If U is a submodule of M^μ, then either $U \supseteq S^\mu$ or $U \subseteq S^{\mu\perp}$.</u>

<u>Proof</u>: Suppose that $u \in U$ and t is a μ-tableau. Then by Corollary 4.7,
$$u\kappa_t = \text{a multiple of } e_t.$$

If we can choose u and t so that this multiple is non-zero, then $e_t \in U$. Since S^μ is generated by e_t, we have $U \supseteq S^\mu$.

If, for every u and t, $u\kappa_t = 0$, then for all u and t
$$0 = \langle u\kappa_t, \{t\}\rangle = \langle u, \{t\}\kappa_t\rangle = \langle u, e_t\rangle \ .$$
That is, $U \subseteq S^{\mu\perp}$.

4.9 THEOREM <u>$S^\mu/(S^\mu \cap S^{\mu\perp})$ is zero or absolutely irreducible. Further if this is non-zero, then $S^\mu \cap S^{\mu\perp}$ is the unique maximal submodule of S^μ, and $S^\mu/(S^\mu \cap S^{\mu\perp})$ is self-dual.</u>

<u>Proof</u>: By the Submodule Theorem, any submodule of S^μ is either S^μ itself, or is contained in $S^\mu \cap S^{\mu\perp}$. Using 1.5, all parts of the Theorem follow at once, except that we have still to prove that $S^\mu/(S^\mu \cap S^{\mu\perp})$ remains irreducible when we extend the field.

Choose a basis $e_1,\ldots,e_k$ for S^μ where each e_i is a polytabloid. (We shall see later how to do this in a special way.) By Theorem 1.6, $\dim(S^\mu/S^\mu \cap S^{\mu\perp})$ is the rank of the Gram matrix with respect to this basis. But the Gram matrix has entries from the prime subfield of F, since the coefficients of tabloids involved in a polytabloid are all ± 1. Therefore, the rank of the Gram matrix is the same over F as over the prime subfield, and so $S^\mu \cap S^{\mu\perp}$ does not increase in dimension if we extend F. Since $S^\mu/(S^\mu \cap S^{\mu\perp})$ is always irreducible, it follows that it is absolutely irreducible.

<u>Remark</u> We shall show that all the irreducible representations of $\mathfrak{S}_n$ turn up as $S^\mu/(S^\mu \cap S^{\mu\perp})$; the Theorem means that we can work over $\mathbb{Q}$ or

the field of p elements. We now concentrate on completing the case where char F = 0, although the remainder of this section also follows from the more subtle approach in section 11. The reader impatient for the more general result can go immediately to sections 10 and 11.

4.10 LEMMA <u>If Θ is an $F\mathfrak{S}_n$-homomorphism from M^λ into M^μ and $S^\lambda \not\subseteq$ Ker Θ, then $\lambda \trianglerighteq \mu$. If $\lambda = \mu$, the restriction of Θ to S^λ is multiplication by a constant.</u>

<u>Remark</u> Ker $\Theta \subseteq S^{\lambda\perp}$ by the Submodule Theorem, since Ker $\Theta \not\supseteq S^\lambda$. The Lemma will later be improved in several ways (cf. 11.3 and 13.17).

<u>Proof:</u> Suppose that t is a λ-tableau. Since $e_t \notin$ Ker Θ,

$$0 \neq e_t\,\Theta = \{t\}\kappa_t\,\Theta = \{t\}\Theta\,\kappa_t$$

$$= (\text{a linear combination of } \mu\text{-tabloids})\kappa_t.$$

By Lemma 4.6, $\lambda \trianglerighteq \mu$, and if $\lambda = \mu$, then $e_t\Theta$ is a multiple of e_t.

4.11 COROLLARY <u>If char F = 0, and Θ is a non-zero element of $\mathrm{Hom}_{F\mathfrak{S}_n}(S^\lambda, M^\mu)$, then $\lambda \trianglerighteq \mu$. If $\lambda = \mu$, then Θ is multiplication by a constant.</u>

<u>Proof:</u> When $F = \mathbb{Q}$, $\langle\ ,\ \rangle$ is an inner product. The rank of the Gram matrix with respect to a basis of S^λ therefore equals dim S^λ for any field of characteristic 0. Thus

<u>when char F = 0, $S^\lambda \cap S^{\lambda\perp} = 0$ and $M^\lambda = S^\lambda \oplus S^{\lambda\perp}$</u>.

Any homomorphism defined on S^λ can therefore be extended to be defined on M^λ by letting it be zero on $S^{\lambda\perp}$. Now apply the Lemma.

4.12 THEOREM (THE ORDINARY IRREDUCIBLE REPRESENTATIONS OF $\mathfrak{S}_n$). <u>The Specht modules over $\mathbb{Q}$ are self-dual and absolutely irreducible, and give all the ordinary irreducible representations of $\mathfrak{S}_n$.</u>

<u>Proof:</u> If $S^\lambda_{\mathbb{Q}} \cong S^\mu_{\mathbb{Q}}$, then $\lambda \trianglerighteq \mu$ by Corollary 4.11. Similarly, $\mu \trianglerighteq \lambda$ so $\lambda = \mu$. Since $S^\lambda_{\mathbb{Q}} \cap S^{\lambda\perp}_{\mathbb{Q}} = 0$, the Theorem follows from Theorem 4.9 and 2.4.

Since M^μ is completely reducible when char F = 0, Corollary 4.11 also gives

4.13 THEOREM <u>If char F = 0, the composition factors of M^μ are S^μ (once) and some of $\{S^\lambda \mid \lambda \triangleright \mu\}$ (possibly with repeats).</u>

Some authors prefer to work inside the group algebra of $\mathfrak{S}_n$, and so we explain how to find a right ideal of the group algebra of $\mathfrak{S}_n$ corresponding to the Specht module.

<u>Given a μ-tableau t</u>, let $\rho_t = \sum_{\sigma \in R_t} \sigma$, so that $\rho_t \in F\mathfrak{S}_n$, and let

$$\Theta\colon \rho_t\,\pi \to \{t\}\pi \qquad (\pi \in \mathfrak{S}_n).$$

This is clearly a well-defined $F\mathfrak{S}_n$ isomorphism from the right ideal $\rho_t F\mathfrak{S}_n$ onto M^μ (It is well-defined, since $\rho_t \pi = \rho_t \Leftrightarrow \pi \in R_t \Leftrightarrow \{t\}\pi = \{t\}$.) Restricting θ to the right ideal $\rho_t \kappa_t F\mathfrak{S}_n$ gives an isomorphism from $\rho_t \kappa_t F\mathfrak{S}_n$ onto S^μ. Using this isomorphism, every result can be interpreted in terms of the group algebra. We prefer the Specht module approach for two reasons. First, the Specht module S^μ depends only on the partition μ, whereas the right ideal $\rho_t \kappa_t F\mathfrak{S}_n$ depends on the particular μ-tableau t. Perhaps more important is that in place of ρ_t, which is a long sum of group elements, we have a single object $\{t\}$; this greatly simplifies manipulations with particular examples, as will be seen in the next section, where we pause in the development to work through some examples illustrating many salient points.

5. EXAMPLES

5.1 EXAMPLE   Reverting to the notation of Example 2.5, where the first row of the tabloids in $M^{(n-1,1)}$ is omitted, we have

$$S^{(n-1,1)} = (\bar{2} - \bar{1})F\mathfrak{S}_n = \{\sum a_i \ \bar{i} \mid a_i \in F,\ a_1 + \ldots + a_n = 0\}$$

$$S^{(n-1,1)\perp} = Sp(\bar{1} + \bar{2} + \ldots + \bar{n}).$$

Clearly, $S^{(n-1,1)\perp} \subseteq S^{(n-1,1)}$ if and only if char F divides n. By the Submodule Theorem

$$0 \subset S^{(1^2)\perp} = S^{(1^2)} \subset M^{(1^2)} \text{ if char } F = 2 \text{ and } n = 2$$

$$0 \subset S^{(n-1,1)\perp} \subset S^{(n-1,1)} \subset M^{(n-1,1)} \text{ if char } F \text{ divides } n > 2$$

are the unique composition series for $M^{(n-1,1)}$ if char F divides n.

The same Theorem shows that when char F does not divide n, $S^{(n-1,1)}$ is irreducible and $M^{(n-1,1)} = S^{(n-1,1)} \oplus S^{(n-1,1)\perp}$.

Note that in all cases $S^{(n-1,1)\perp} \cong S^{(n)}$ and $\dim S^{(n-1,1)} = n-1$.

5.2 EXAMPLE   We examine $M^{(3,2)}$ in detail. A (3,2)-tabloid is determined by the unordered pair of numbers $\overline{ij}$ which make up its second row. To get a geometric picture of $M^{(3,2)}$, consider the set of graphs (without loops) on 5 points, where we allow an edge to be "weighted" by a field coefficient. By identifying $\overline{ij}$ with the edge joining point i to point j, we have constructed an isomorphic copy of $M^{(3,2)}$. For example,

$$\frac{\overline{2\ 5\ 1}}{\underline{3\ 4}} - \frac{\overline{3\ 5\ 1}}{\underline{2\ 4}} - \frac{\overline{2\ 4\ 1}}{\underline{3\ 5}} + \frac{\overline{3\ 4\ 1}}{\underline{2\ 5}} \quad \text{corresponds to}$$

Any "quadrilateral with alternate edges weighted ± 1" is a generator for the Specht module $S^{(3,2)}$.

Let t_1, t_2, t_3, t_4, t_5 =
$\begin{matrix}1&3&5\\2&4\end{matrix}$ $\begin{matrix}1&2&5\\3&4\end{matrix}$ $\begin{matrix}1&3&4\\2&5\end{matrix}$ $\begin{matrix}1&2&4\\3&5\end{matrix}$ $\begin{matrix}1&2&3\\4&5\end{matrix}$

respectively. Then $e_{t_1}, \ldots, e_{t_5}$ correspond to

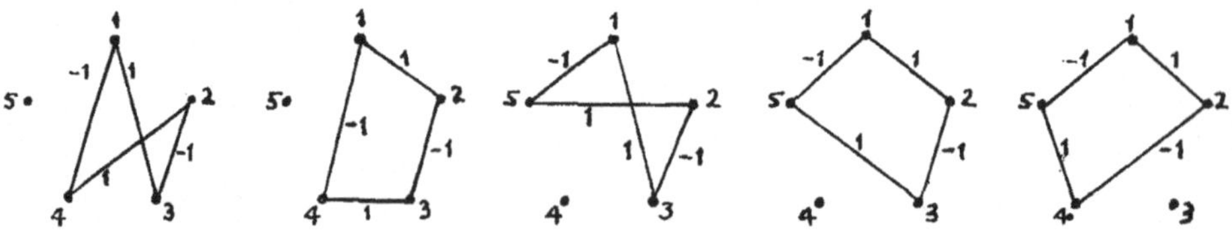

respectively.

The 10 edges are ordered by 3.10:

$$\overline{1\ 2} < \overline{1\ 3} < \overline{2\ 3} < \overline{1\ 4} < \overline{2\ 4} < \overline{3\ 4} < \overline{1\ 5} < \overline{2\ 5} < \overline{3\ 5} < \overline{4\ 5}\ .$$

The last edges involved in $e_{t_1}, \ldots, e_{t_5}$ are $\overline{2\ 4}, \overline{3\ 4}, \overline{2\ 5}, \overline{3\ 5}, \overline{4\ 5}$

(which correspond to $\{t_1\},\ldots,\{t_5\}$.) Since these last edges are different, $e_{t_1},\ldots,e_{t_5}$ are linearly independent. Note that it is far from clear that they also span the Specht module, but we shall prove this later. Assuming that they do give a basis, the Gram matrix with respect to this basis is

$$A = \begin{pmatrix} 4 & 2 & 2 & 1 & -1 \\ 2 & 4 & 1 & 2 & 1 \\ 2 & 1 & 4 & 2 & 1 \\ 1 & 2 & 2 & 4 & 2 \\ -1 & 1 & 1 & 2 & 4 \end{pmatrix}$$

One checks that if char F = 0 or char F ≥ 5, rank A = 5
if char F = 3, rank A = 1
if char F = 2, rank A = 4.

Therefore, $\dim(S^{(3,2)}/S^{(3,2)} \cap S^{(3,2)\perp}) = 5$ unless char F = 2 or 3, when the dimension is 4 or 1, respectively.

Let us find $S^{(3,2)\perp}$. Certainly,

$\Gamma =$ 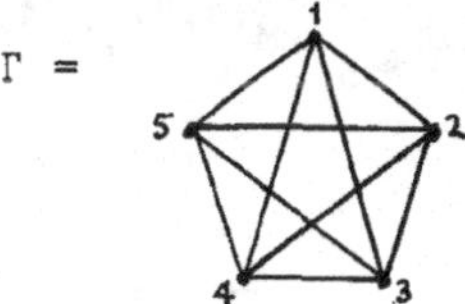

and 5 graphs like $\Gamma(-1) =$ 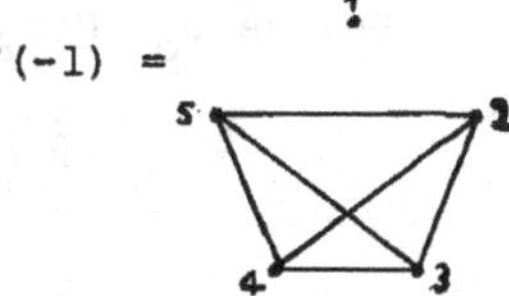

are orthogonal to "quadrilaterals with alternate edges weighted ± 1". (An unlabelled edge is assumed to have weight 1). That is, they belong to $S^{(3,2)\perp}$. ($\Gamma(-i)$ is defined by $\Gamma(-i) = \Gamma(-1)(1\ i)$ for $1 \le i \le 5$.)

Now, $\Gamma(-1) + \Gamma(-2) + \ldots + \Gamma(-5) = 3\Gamma$. It is easy to verify that $\Gamma(-1),\ldots,\Gamma(-5)$ are linearly independent if char F ≠ 3, and that they span a space of dimension 4 when char F = 3. Hence

$S^{(3,2)\perp}$ is spanned by $\Gamma, \Gamma(-1), \Gamma(-2), \ldots, \Gamma(-5)$

since $S^{(3,2)\perp}$ has dimension 5 (by 1.3).

When <u>char F = 2</u>, $e_{t_2} + e_{t_3} + e_{t_4} + e_{t_5} = \Gamma$. Therefore, $\Gamma \in S^{(3,2)} \cap S^{(3,2)\perp}$ in this case, and by dimensions it spans $S^{(3,2)} \cap S^{(3,2)\perp}$.

When <u>char F = 3</u>, $e_{t_1} + e_{t_2} = \Gamma(-5)$, and now $\Gamma(-1),\ldots,\Gamma(-5)$ span $S^{(3,2)} \cap S^{(3,2)\perp}$.

We do not yet have a convenient way of checking whether or not a graph belongs to $S^{(3,2)}$. However, every such graph certainly satisfies the two conditions:

<u>5.3</u> (i) The sum of the coefficients of the edges is zero.

(ii) The valency of each point is zero. (Formally: the sum of the coefficients of the edges at each point is zero.)

These conditions hold because a generator for $S^{(3,2)}$ satisfies the conditions. In fact, the properties characterize $S^{(3,2)}$ and enable us rapidly to check that $\Gamma \in S^{(3,2)}$ when char $F = 2$ (Γ has an even number of edges, and each point has even valency), and that $\Gamma(-5) \in S^{(3,2)}$ when char $F = 3$ ($\Gamma(-5)$ has 6 edges and each point has valency 0 or 3).

So far, we have highlighted two problems to be discussed later:

(a) Find a basis for the general Specht module like that given above. (N.B. It is not obvious even that dim S^{μ} is independent of the field.)

(b) Find conditions similar to 5.3 characterizing the Specht module as a submodule of M^{μ}(cf. the second expression for $S^{(n-1,1)}$ in Example 5.1).

We have proved that $e_{t_1},\ldots,e_{t_5}$ are linearly independent; here, as in the general case, it is a lot harder to prove that they span $S^{(3,2)}$. This example is concluded by a simultaneous proof that $e_{t_1},\ldots,e_{t_5}$ form a basis of $S^{(3,2)}$ and that conditions 5.3 characterize $S^{(3,2)}$.

Define $\psi_0 \in \mathrm{Hom}_{F\mathfrak{S}_5}(M^{(3,2)},M^{(5)})$ and $\psi_1 \in \mathrm{Hom}_{F\mathfrak{S}_5}(M^{(3,2)},M^{(4,1)})$ by

$$\psi_0 : \begin{array}{l}\overline{\underline{a\ b\ c}}\\ \underline{d\ e}\end{array} \rightarrow \overline{\underline{a\ b\ c\ d\ e}}$$

$$\psi_1 : \begin{array}{l}\overline{\underline{a\ b\ c}}\\ \underline{d\ e}\end{array} \rightarrow \begin{array}{l}\overline{\underline{a\ b\ c\ e}}\\ \underline{d}\end{array} + \begin{array}{l}\overline{\underline{a\ b\ c\ d}}\\ \underline{e}\end{array} \quad (\text{i.e. } \overline{de} \rightarrow \bar{d} + \bar{e})$$

Now, conditions 5.3(i) or (ii) hold for an element v of $M^{(3,2)}$ if and only if $v \in \mathrm{Ker}\,\psi_0$ or $v \in \mathrm{Ker}\,\psi_1$, repectively. Therefore $S^{(3,2)} \subseteq \mathrm{Ker}\,\psi_0 \cap \mathrm{Ker}\,\psi_1$ (cf. Lemma 4.10), and we want to prove equality.

Write $S^{(3,1),(3,2)}$ for the space spanned by graphs of the form

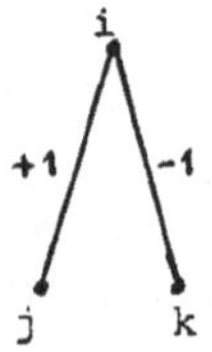

i

+1 -1

j k

$$= \overline{i\ j} - \overline{i\ k}$$

Now, $S^{(3,1),(3,2)} \subseteq \mathrm{Ker}\,\psi_0$ and ψ_1 sends $S^{(3,1),(3,2)}$ onto $S^{(4,1)}$ (since $\psi_1 : \overline{ij} - \overline{ik} \rightarrow \bar{i}+\bar{j}-\bar{i} - \bar{k} = \bar{j} - \bar{k}$). Therefore, we have the following series for $M^{(3,2)}$:

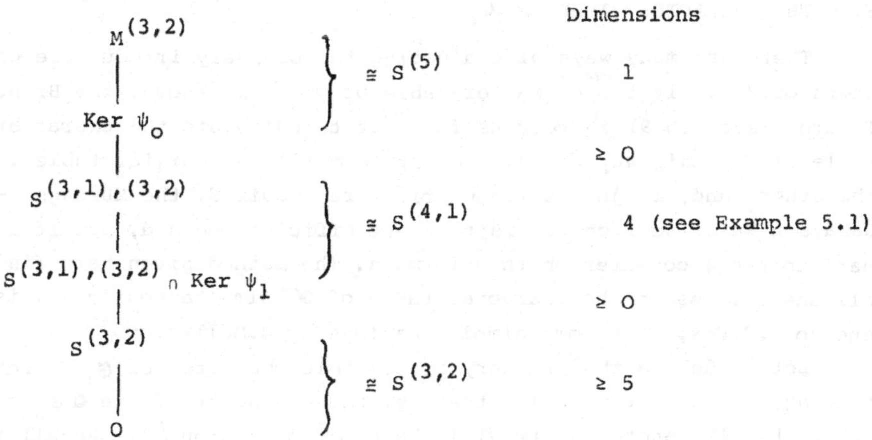

But dim $M^{(3,2)} = 10$, so we have equality in all possible places. In particular, dim $S^{(3,2)} = 5$ and $S^{(3,2)} = \text{Ker } \psi_o \cap \text{Ker } \psi_1$, as we wished to prove.

5.4 EXAMPLE $S^{(2,2)}$ is spanned by the graphs

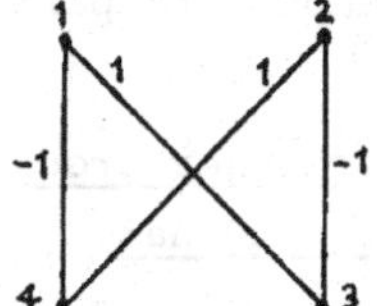

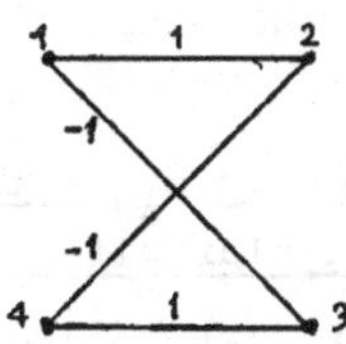

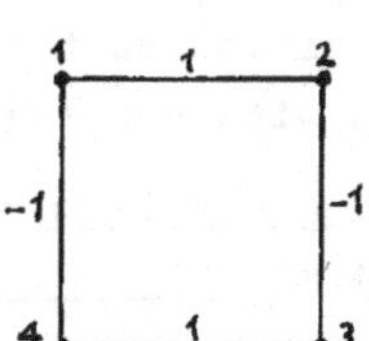

Clearly, the first two form a basis.

When char F = 2, $S^{(2,2)} \subseteq S^{(2,2)\perp}$. The reason underlying this is that any polytabloid contains none or both edges of the following pairs of edges:

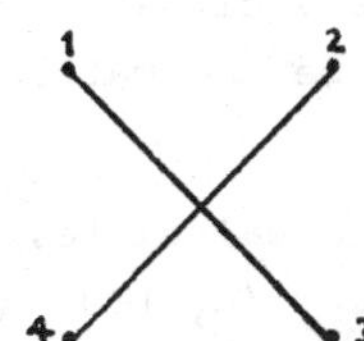

6. THE CHARACTER TABLE OF $\mathfrak{S}_n$

There are many ways of evaluating the ordinary irreducible characters of $\mathfrak{S}_n$. If the character table of $\mathfrak{S}_{n-1}$ is known, the Branching Theorem (section 9) is very useful, but to calculate the character table of $\mathfrak{S}_n$ this way we have to work out all the earlier tables. On the other hand, if just a few entries are required, the Murnaghan-Nakayama Rule (section 21) is the most efficient method, but it is hard to use a computer on this formula. The method given here finds all the entries in the character table of $\mathfrak{S}_n$ simultaneously. It is due to R.F.Fox, with some simplifications by G.Mullineux.

Let χ^λ denote the ordinary irreducible character of $\mathfrak{S}_n$ corresponding to the partition λ - that is, the character of the $\mathbb{Q}\mathfrak{S}_n$ module $S^\lambda_{\mathbb{Q}}$. Let 1_G denote the trivial character of a group G. Recall that $\mathfrak{S}_\lambda$ is a Young subgroup, and that $1_{\mathfrak{S}_\lambda} \uparrow \mathfrak{S}_n$ is the character of $M^\lambda_{\mathbb{Q}}$, by 4.2 (The notation $\uparrow G$ means "induced up to G" and $\downarrow G$ means "restricted to G".)

All the matrices in this section will have rows and columns indexed by partitions of n, in dictionary order (3.4). Since $M^\lambda_{\mathbb{Q}}$ has $S^\lambda_{\mathbb{Q}}$ as a composition factor once, and the other factors correspond to partitions μ with $\mu \rhd \lambda$ (Theorem 4.13),

6.1 <u>The matrix $m = (m_{\lambda\mu})$ given by $m_{\lambda\mu}$ = the character inner product $(1_{\mathfrak{S}_\lambda} \uparrow \mathfrak{S}_n, \chi^\mu)$ is lower triangular with 1's down the diagonal.</u>

(see the example for $\mathfrak{S}_5$, below). It follows at once that the matrix $B = (b_{\lambda\mu})$ given by

$$b_{\lambda\mu} = |\mathfrak{S}_\mu|\ (\chi^\lambda, 1_{\mathfrak{S}_\mu} \uparrow \mathfrak{S}_n)$$

is upper triangular.

Let $\mathcal{C}_\mu$ denote the conjugacy class of $\mathfrak{S}_n$ corresponding to the partition μ, and let $A = (a_{\lambda\mu})$ be the matrix given by

$$a_{\lambda\mu} = |\mathfrak{S}_\lambda \cap \mathcal{C}_\mu|$$

The matrix A is not hard to calculate, and we claim that once it is known, the character table $C = (c_{\lambda\mu})$ of $\mathfrak{S}_n$ can be calculated by straightforward matrix manipulations. First note that

$$\sum_\mu c_{\lambda\mu}\, a_{\nu\mu} = |\mathfrak{S}_\nu|\,(\chi^\lambda \downarrow \mathfrak{S}_\nu, 1_{\mathfrak{S}_\nu}) = b_{\lambda\nu}.$$

Therefore, $B = CA'$, where A' is the transpose of A.

But,
$$\sum_\mu b_{\mu\lambda}\, b_{\mu\nu} = |\mathfrak{S}_\lambda||\mathfrak{S}_\nu|\,(1_{\mathfrak{S}_\lambda}\uparrow \mathfrak{S}_n,\ 1_{\mathfrak{S}_\nu}\uparrow \mathfrak{S}_n)$$
$$= |\mathfrak{S}_\lambda||\mathfrak{S}_\nu|\,(1_{\mathfrak{S}_\lambda}\uparrow \mathfrak{S}_n \downarrow \mathfrak{S}_\nu,\ 1_{\mathfrak{S}_\nu})$$
$$= |\mathfrak{S}_\lambda| \sum_\mu (1_{\mathfrak{S}_\lambda}\uparrow \mathfrak{S}_n \text{ evaluated on an element of type}$$

$\mu) . |\mathfrak{S}_\nu \cap \mathcal{C}_\mu|$

$$= \sum_\mu (n! / |\mathfrak{S}_\mu|) \, |\mathfrak{S}_\lambda \cap \mathcal{C}_\mu| |\mathfrak{S}_\nu \cap \mathcal{C}_\mu|$$

$$= \sum_\mu (n! / |\mathfrak{S}_\mu|) \, a_{\lambda\mu} \, a_{\nu\mu}$$

If A is known, we can solve these equations by starting at the top left hand corner of B, working down each column in turn, and proceeding to the next column on the right. Since B is upper triangular, there is only one unknown to be calculated at each stage, and this can be found, since B has non-negative entries. Therefore

6.2 THEOREM <u>If the matrix $A = (a_{\lambda\mu})$, where $a_{\lambda\mu} = |\mathfrak{S}_\lambda \cap \mathcal{C}_\mu|$ is known, then we can find the unique non-negative upper triangular matrix $B = (b_{\lambda\mu})$ satisfying the equations</u>

$$\sum_\mu b_{\mu\lambda} \, b_{\mu\nu} = \sum_\mu (n! / |\mathfrak{S}_\mu|) a_{\lambda\mu} \, a_{\nu\mu}$$

<u>and the character table C of $\mathfrak{S}_n$ is given by $C = BA'^{-1}$.</u>

6.3 EXAMPLE Suppose n = 5. Then

A =

	(5)	(4,1)	(3,2)	$(3,1^2)$	$(2^2,1)$	$(2,1^3)$	(1^5)
(5)	24	30	20	20	15	10	1
(4,1)		6	0	8	3	6	1
✓ (3,2)			2	2	3	4	1
$(3,1^2)$				2	0	3	1
$(2^2,1)$					1	2	1
$(2,1^3)$						1	1
(1^5)							1

B =

	(5)	(4,1)	(3,2)	$(3,1^2)$	$(2^2,1)$	$(2,1^3)$	(1^5)
(5)	120	24	12	6	4	2	1
(4,1)		24	12	12	8	6	4
(3,2)			12	6	8	6	5
$(3,1^2)$				6	4	6	6
$(2^2,1)$					4	4	5
$(2,1^3)$						2	4
(1^5)							1

$$C = \begin{array}{c} (5) \\ (4,1) \\ (3,2) \\ (3,1^2) \\ (2^2,1) \\ (2,1^3) \\ (1^5) \end{array} \overset{\begin{array}{ccccccc} (5) & (4,1) & (3,2) & (3,1^2) & (2^2,1) & (2,1^3) & (1^5) \end{array}}{\begin{pmatrix} 1 & 1 & 1 & 1 & 1 & 1 & 1 \\ -1 & 0 & -1 & 1 & 0 & 2 & 4 \\ 0 & -1 & 1 & -1 & 1 & 1 & 5 \\ 1 & 0 & 0 & 0 & -2 & 0 & 6 \\ 0 & 1 & -1 & -1 & 1 & -1 & 5 \\ -1 & 0 & 1 & 1 & 0 & -2 & 4 \\ 1 & -1 & -1 & 1 & 1 & -1 & 1 \end{pmatrix}}$$

The columns of the character table are in the reverse order to the usual one - in particular, the degrees of the irreducible characters appear down the last column - because we have chosen to take the dictionary order on both the rows and the columns.

6.4 NOTATION Equations like $[3][2] = [5] + [4,1] + [3,2]$ are to be interpreted as saying that $M_{\mathbb{Q}}^{(3,2)}$ has composition factors isomorphic to $S_{\mathbb{Q}}^{(5)}$, $S_{\mathbb{Q}}^{(4,1)}$ and $S_{\mathbb{Q}}^{(3,2)}$. In general if λ is a partition of n,

$$[\lambda_1][\lambda_2][\lambda_3]\ldots = \sum_{\mu} m_{\lambda\mu}\ [\mu]$$

means that $M_{\mathbb{Q}}^{\lambda}$ has $S_{\mathbb{Q}}^{\mu}$ as a factor with multiplicity $m_{\lambda\mu}$. ($m = (m_{\lambda\mu})$ is the matrix defined in 6.1).

By dividing each column of the matrix B by the number at the top of that column (which equals $|\mathfrak{S}_\mu|$), and transposing, the matrix m is obtained. In the above example,

$$m = \begin{array}{c} [5] \\ [4][1] \\ [3][2] \\ [3][1]^2 \\ [2]^2[1] \\ [2][1]^3 \\ [1]^5 \end{array} \overset{\begin{array}{ccccccc} [5] & [4,1] & [3,2] & [3,1^2] & [2^2,1] & [2,1^3] & [1^5] \end{array}}{\begin{pmatrix} 1 & & & & & & \\ 1 & 1 & & & & & \\ 1 & 1 & 1 & & & & \\ 1 & 2 & 1 & 1 & & & \\ 1 & 2 & 2 & 1 & 1 & & \\ 1 & 3 & 3 & 3 & 2 & 1 & \\ 1 & 4 & 5 & 6 & 5 & 4 & 1 \end{pmatrix}}$$

Notice that the results $[4][1] = [5] + [4,1]$ and $[3][2] = [5] + [4,1] + [3,2]$ are in agreement with Examples 5.1 and 5.2. Young's Rule in section 14 shows how to evaluate the matrix m directly.

Theorem 6.2 has the interesting

6.5 COROLLARY <u>The determinant of the character table of $\mathfrak{S}_n$ is the product of all the parts of all the partitions of n.</u>

<u>Proof</u>: $a_{\lambda\lambda} = \prod_i (\lambda_i - 1)!$ and $b_{\lambda\lambda} = |\mathfrak{S}_\lambda| = \prod_i \lambda_i!$

Since A and B are upper triangular and $B = CA'$, we have

det C = $\prod_{\lambda} \prod_{i} \lambda_i$, as claimed.

Recall that the partition λ' conjugate to λ is obtained by "turning λ on its side" (see definition 3.5). The character table of $\mathfrak{S}_5$ in Example 6.3 exhibits the property:

<u>6.6</u> $\underline{\chi^{\lambda'} = \chi^{\lambda} \otimes \chi^{(1^n)}}$

We prove this in general by showing

6.7 THEOREM <u>$S^{\lambda}_{\mathbb{Q}} \otimes S^{(1^n)}_{\mathbb{Q}}$ is isomorphic to the dual of $S^{\lambda'}_{\mathbb{Q}}$</u>.

<u>Remark</u> Since $S^{\lambda'}_{\mathbb{Q}}$ is self-dual, we may omit the words "the dual of" from the statement of the Theorem, but we shall later prove the analogous Theorem over an arbitrary field, where the distinction between $S^{\lambda'}$ and its dual must be made.

<u>Proof</u>: Let t be a <u>given λ-tableau</u>, and let t' be the corresponding λ' tableau.

e.g. if t = 1 2 3 then t'= 1 4
4 5 2 5
3

Let $\rho_{t'} = \sum\{\pi \mid \pi \in R_{t'}\}$ and $\kappa_{t'} = \sum\{(\mathrm{sgn}\ \pi)\pi \mid \pi \in C_{t'}\}$, as usual. Let u be a generator for $S^{(1^n)}_{\mathbb{Q}}$, so that $u\pi = (\mathrm{sgn}\ \pi)u$ when $\pi \in \mathfrak{S}_n$.

It is routine to verify that there is a well-defined $\mathbb{Q}\mathfrak{S}_n$-epimorphism Θ from $M^{\lambda'}_{\mathbb{Q}}$ onto $S^{\lambda}_{\mathbb{Q}} \otimes S^{(1^n)}_{\mathbb{Q}}$ sending $\{t'\}$ to $(\{t\} \otimes u)\rho_{t'}$; Θ is given by

<u>6.8</u> $\Theta\colon \{t'\pi\} \to (\{t\} \otimes u)\rho_{t'}\pi = (\{t\}\kappa_t \otimes u)\pi = (\mathrm{sgn}\ \pi)\{t\pi\}\kappa_{t\pi} \otimes u.$

Θ sends $\{t'\}\kappa_{t'}$ to $(\{t'\} \otimes u)\rho_{t'}\kappa_{t'} = \{t\}\kappa_t\rho_t \otimes u.$

Now, $\langle\{t\}\kappa_t\rho_t, \{t\}\rangle = \langle\{t\}\kappa_t, \{t\}\rho_t\rangle$
$= \langle\{t\}\kappa_t, |R_t|\{t\}\rangle = |R_t|.$

Since $|R_t|$ is a non-zero element of $\mathbb{Q}$, $\{t'\}\kappa_{t'}\Theta \neq 0$. Thus Ker $\Theta \not\supseteq S^{\lambda'}_{\mathbb{Q}}$, and, by the Submodule Theorem, Ker $\Theta \subseteq S^{\lambda'\perp}_{\mathbb{Q}}$. Therefore, $\dim S^{\lambda}_{\mathbb{Q}} = \dim \mathrm{Im}\,\Theta = \dim(M^{\lambda'}_{\mathbb{Q}}/\mathrm{Ker}\,\Theta) \geq \dim(M^{\lambda'}_{\mathbb{Q}}/S^{\lambda'\perp}_{\mathbb{Q}}) = \dim S^{\lambda'}_{\mathbb{Q}}$ (*).

Similarly, $\dim S^{\lambda'}_{\mathbb{Q}} \geq \dim S^{\lambda''}_{\mathbb{Q}} = \dim S^{\lambda}_{\mathbb{Q}}$. Therefore, $\dim S^{\lambda}_{\mathbb{Q}} = \dim S^{\lambda'}_{\mathbb{Q}}$ and we have equality in (*). Thus, Ker $\Theta = S^{\lambda\perp}_{\mathbb{Q}}$. The theorem is now proved, since we have constructed an isomorphism between $M^{\lambda'}_{\mathbb{Q}}/S^{\lambda'\perp}_{\mathbb{Q}}$ ($\cong$ dual of $S^{\lambda'}_{\mathbb{Q}}$, by 1.4) and $S^{\lambda}_{\mathbb{Q}} \otimes S^{(1^n)}_{\mathbb{Q}}$.

<u>Remark</u> Corollary 8.5 will give $\dim S^{\lambda} = \dim S^{\lambda'}$, trivially, but this shortens the proof by only one line.

There is one non-trivial character of $\mathfrak{S}_n$ which can always be evaluated quickly, namely $\chi^{(n-1,1)}$:

6.9 LEMMA <u>The value of $\chi^{(n-1,1)}$ on a permutation π is one less than the number of fixed points of π.</u>

<u>Proof</u>: The trace of π, acting on the permutation module $M^{(n-1,1)}$, is clearly the number of fixed points of π. Since

$$M_{\mathbb{Q}}^{(n-1,1)} \cong S_{\mathbb{Q}}^{(n)} \oplus S_{\mathbb{Q}}^{(n-1,1)}$$

(cf. Example 5.1), the result follows at once.

We can thus write down four characters, $\chi^{(n)}$, $\chi^{(n-1,1)}$, $\chi^{(1^n)}$ and $\chi^{(2,1^{n-2})}$ $(= \chi^{(n-1,1)} \otimes \chi^{(1^n)})$ of $\mathfrak{S}_n$ at once. The best way of finding the character table of $\mathfrak{S}_n$ for small n is to deduce the remaining characters from these, using the column orthogonality relations.

7. THE GARNIR RELATIONS

For this section, <u>let t be a given μ-tableau</u>. We want to find elements of the group algebra of $\mathfrak{S}_n$ which annihilate the given polytabloid e_t.

Let X be a subset of the ith column of t, and Y be a subset of the (i + 1)th column of t.

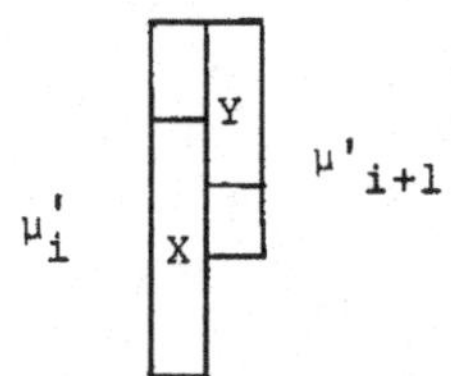

Let $\sigma_1, \ldots, \sigma_k$ be coset representatives for $\mathfrak{S}_X \times \mathfrak{S}_Y$ in $\mathfrak{S}_{X \cup Y}$, and let $G_{X,Y} = \sum_{j=1}^{k} (\text{sgn } \sigma_j)\sigma_j$. $G_{X,Y}$ is called a <u>Garnir element</u> . (Garnir [5]).

In all applications, X will be taken at the end of the ith column of t and Y will be at the beginning of the (i+1)th column. The permutations $\sigma_1, \ldots, \sigma_k$ are, of course, not unique, but for practical purposes note that we may take $\sigma_1, \ldots, \sigma_k$ so that $t\sigma_1, t\sigma_2, \ldots, t\sigma_k$ are all the tableaux which agree with t except in the positions occupied by $X \cup Y$, and whose entries increase vertically downwards in the positions occupied by $X \cup Y$.

7.1 EXAMPLE if $t = \begin{matrix} 1 & 2 \\ 4 & 3 \\ 5 & \end{matrix}$, X = {4,5} and Y = {2,3} then $t\sigma_1, \ldots, t\sigma_k$ may be taken as

$$t = t_1 = \begin{matrix} 1 & 2 \\ 4 & 3 \\ 5 & \end{matrix} \quad t_2 = \begin{matrix} 1 & 2 \\ 3 & 4 \\ 5 & \end{matrix} \quad t_3 = \begin{matrix} 1 & 2 \\ 3 & 5 \\ 4 & \end{matrix} \quad t_4 = \begin{matrix} 1 & 3 \\ 2 & 4 \\ 5 & \end{matrix} \quad t_5 = \begin{matrix} 1 & 3 \\ 2 & 5 \\ 4 & \end{matrix} \quad t_6 = \begin{matrix} 1 & 4 \\ 2 & 5 \\ 3 & \end{matrix}$$

when sgn $\sigma_i = 1$ for i = 1,3,4,6, sgn $\sigma_i = -1$ for i = 2,5 and $G_{X,Y} = 1 - (3\ 4) + (3\ 5\ 4) + (2\ 3\ 4) - (2\ 3\ 5\ 4) + (2\ 4)(3\ 5)$.

7.2 THEOREM <u>If $|X \cup Y| > \mu'_i$, then $e_t G_{X,Y} = 0$ (for any base field).</u>

<u>Proof</u>: (See Peel [19]) Write $\mathfrak{S}^-_X \mathfrak{S}^-_Y$ for $\sum\{(\text{sgn } \sigma)\sigma \mid \sigma \in \mathfrak{S}_X \times \mathfrak{S}_Y\}$

and $\mathfrak{S}^-_{X\cup Y}$ for $\sum\{(\text{sgn } \sigma)\sigma \mid \sigma \in \mathfrak{S}_{X \cup Y}\}$

Since $|X \cup Y| > \mu'_i$, for every τ in the column stabilizer of t, some pair of numbers in $X \cup Y$ are in the same row of $t\tau$. Hence, in the usual way, $\{t\tau\}\mathfrak{S}^-_{X\cup Y} = 0$. Therefore, $\{t\}\kappa_t \mathfrak{S}^-_{X\cup Y} = 0$.

Now, $\mathfrak{S}^-_X \mathfrak{S}^-_Y$ is a factor of κ_t, and $\mathfrak{S}^-_{X\cup Y} = \mathfrak{S}^-_X \mathfrak{S}^-_Y G_{X,Y}$.

Therefore $\quad 0 = \{t\}\kappa_t \, G^-_{X \cup Y} = |X|!|Y|!\{t\}\kappa_t \, G_{X,Y}$

Thus, $\{t\}\kappa_t \, G_{X,Y} = 0$ when the base field is $\mathbb{Q}$, and since all the tabloid coefficients here are integers, the same holds over any field.

7.3 EXAMPLE Referring to Example 7.1, we have

$$0 = e_t \, G_{X,Y} = e_{t_1} - e_{t_2} + e_{t_3} + e_{t_4} - e_{t_5} + e_{t_6}$$

so $e_t = e_{t_2} - e_{t_3} - e_{t_4} + e_{t_5} - e_{t_6}$.

8. THE STANDARD BASIS OF THE SPECHT MODULE

8.1 DEFINITIONS t is a standard tableau if the numbers increase along the rows and down the columns of t. $\{t\}$ is a standard tabloid if there is a standard tableau in the equivalence class $\{t\}$. e_t is a standard polytabloid if t is standard.

In Example 5.2, the 5 standard (3,2)-tableaux and the corresponding standard polytabloids are listed.

A standard tabloid contains a unique standard tableau, since the numbers have to increase along the rows of a standard tableau. It is annoying that a polytabloid may involve more than one standard tabloid (In Example 5.2, e_{t_5} involves $\overline{4\ 5}$ and $\overline{2\ 4}$).

We prove that the standard polytabloids form a basis for the Specht module, defined over any field.

The μ-tabloids have been totally ordered by definition 3.10. The linear independence of the standard polytabloids follows from the trivial

8.2 LEMMA Suppose that $v_1, v_2, \ldots, v_m$ are elements of M^μ and that $\{t_i\}$ is the last tabloid involved in v_i. If the tabloids $\{t_i\}$ are all different, then $v_1, v_2, \ldots, v_m$ are linearly independent.

Proof: We may assume that $\{t_1\} < \{t_2\} < \ldots < \{t_m\}$. If $a_1v_1 + \ldots + a_mv_m = 0$ $(a_i \in F)$ and $a_{j+1} = \ldots = a_m = 0$, then $a_j = 0$, since $\{t_j\}$ is involved in v_j and in no v_k with $k < j$. Therefore, $a_1 = \ldots = a_m = 0$.

It is clear that $\{t\}$ is the last tabloid involved in e_t when t is standard, and this is all we need to deduce that the standard polytabloids are linearly independent, but we go for a stronger result, using the partial order (3.11) on tabloids:

8.3 LEMMA If t has numbers increasing down columns, then all the tabloids $\{t'\}$ involved in e_t satisfy $\{t'\} \trianglelefteq \{t\}$.

Proof: If $t' = t\pi$ with π a non-identity element of the column stabilizer of t, then in some column of t' there are numbers $w < x$ with w lower than x. Thus, by 3.15, $\{t'\} \triangleleft \{t'(wx)\}$. Since $\{t'(w\,x)\}$ is involved in e_t, induction shows that $\{t'(w\,x)\} \trianglelefteq \{t\}$. Therefore, $\{t'\} \triangleleft \{t\}$.

8.4 THEOREM $\{e_t \mid t$ is a standard μ-tableau$\}$ is a basis for S^μ.

Proof: (See Peel [19]) We have already proved that the standard polytabloids are linearly independent, and we now use the Garnir relations to prove that any polytabloid can be written as a linear combination of standard polytabloids - a glance at Example 7.3 should show the reader how to do this.

First we write $[t]$ for the column equivalence class of t; that is $[t] = \{t_1 | t_1 = t\pi \text{ for some } \pi \in C_t\}$. The column equivalence classes are totally ordered in a way similar to the order 3.10 on the row equivalence classes.

Suppose that t is not standard. By induction, we may assume that $e_{t'}$ can be written as a linear combination of standard polytabloids when $[t'] < [t]$ and prove the same result for e_t. Since $e_t\pi = (\text{sgn}\,\pi)e_t$ when $\pi \in C_t$, we may suppose that the entries in t are in increasing order down columns. Unless t is standard, some adjacent pair of columns, say the jth and (j+1)th columns, have entries $a_1 < a_2 < \ldots < a_r$, $b_1 < b_2 < \ldots < b_s$ with $a_q > b_q$ for some q

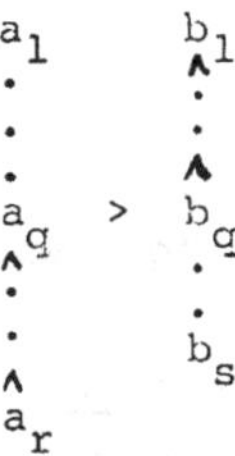

Let $X = \{a_q,\ldots,a_r\}$ and $Y = \{b_1,\ldots,b_q\}$ and consider the corresponding Garnir element $G_{X,Y} = \sum(\text{sgn}\,\sigma)\sigma$, say. By Theorem 7.2

$$0 = e_t \sum(\text{sgn}\,\sigma)\sigma = \sum(\text{sgn}\,\sigma)e_{t\sigma} .$$

Because $b_1 < \ldots < b_q < a_q < \ldots < a_r$, $[t\sigma] < [t]$ for $\sigma \neq 1$. Since $e_t = -\sum_{\sigma \neq 1} (\text{sgn}\,\sigma)e_{t\sigma}$, the result follows from our induction hypothesis.

8.5 COROLLARY <u>The dimension of the Specht module S^μ is independent of the ground field, and equals the number of standard μ-tableaux.</u>

<u>Remark</u> An independent proof of Theorem 8.4 is given in section 17.

8.6 COROLLARY <u>In $S^\mu_{\mathbb{Q}}$ any polytabloid can be written as an integral linear combination of standard polytabloids.</u>

<u>Proof:</u> This result comes from the proof of Theorem 8.4; alternatively, see 8.9 below.

8.7 COROLLARY <u>The matrices representing $\mathfrak{S}_n$ over $\mathbb{Q}$ with respect to the standard basis of $S^\mu_{\mathbb{Q}}$ all have integer coefficients.</u>

<u>Proof:</u> $e_t\pi = e_{t\pi}$. Now apply Corollary 8.6.

8.8 COROLLARY <u>If v is a non-zero element of S^μ, then every last tabloid (in the partial order $\trianglelefteq$ on tabloids) involved in v is standard.</u>

<u>Proof:</u> Since v is a linear combination of standard polytabloids, the result follows from Lemma 8.3.

8.9 COROLLARY <u>If $v \in S^{\mu}_{\mathbb{Q}}$ and the coefficients of the tabloids involved in v are all integers, then v is an integral linear combination of standard polytabloids</u>.

<u>Proof</u>: We may assume that v is non-zero. Let $\{t\}$ be the last (in the < order) tabloid involved in v, with coefficient $a \in \mathbb{Z}$, say. By the last corollary, $\{t\}$ is standard. Now Lemma 8.3 shows that the last tabloid in $v - a\,e_t$ is before $\{t\}$, so by induction $v - a\,e_t$ is an integral linear combination of standard polytabloids. Therefore, the same is true of v.

8.10 COROLLARY <u>If $v \in S^{\mu}_{\mathbb{Q}}$ and the coefficients of the tabloids involved in v are all integers, then we may reduce all these integers modulo p and obtain an element S^{μ}_{F}, where F is the field of p elements</u>.

<u>Proof</u>: By the last Corollary, v is an integral linear combination of standard polytabloids, $v = \sum a_i\, e_i$, say $(a_i \in \mathbb{Z})$. Reducing modulo p all the tabloid coefficients in v, we obtain $\bar{v}$, say. Let $\bar{a}_i$ be a_i modulo p. The equation $\bar{v} = \sum \bar{a}_i\, e_i$ shows that $\bar{v} \in S^{\mu}_{F}$.

<u>Remark</u> If we knew only that the standard polytabloids span $S^{\mu}_{\mathbb{Q}}$, the proof of Corollary 8.10 shows that any polytabloid can be written as a linear combination of standard polytabloids over any field. Therefore, we can deduce that the standard polytabloids span S^{μ} over any field, knowing only the same information over $\mathbb{Q}$.

8.11 COROLLARY <u>If F is the field of p elements, then S^{μ}_{F} is the p-modular representation of $\mathfrak{S}_n$ obtained from $S^{\mu}_{\mathbb{Q}}$</u>.

<u>Proof</u>: Apply the last Corollary.

8.12 COROLLARY <u>There is a basis of S^{μ}, all of whose elements involve a unique standard tabloid.</u>

<u>Proof</u>: Let $\{t_1\} < \{t_2\} < \ldots$ be the standard μ-tabloids. $\{t_1\}$ is the only standard tabloid involved in e_{t_1} by Lemma 8.3. e_{t_2} may involve $\{t_1\}$, with coefficient a, say. Replace e_{t_2} by $f_{t_2} = e_{t_2} - a\,e_{t_1}$. Then $\{t_2\}$ is the only standard tabloid involved in f_{t_2}. Continuing in this fashion, we construct the desired basis.

Corollary 8.12 is useful in numerical calculations.

8.13 EXAMPLE Taking $e_{t_1},\ldots,e_{t_5}$ as in Example 5.2, each involves just one standard tabloid, except e_{t_5} which involves $\overline{2\ 4}$ as well as $\overline{4\ 5}$. Replace e_{t_5} by $f_{t_5} = e_{t_1} + e_{t_5}$. Then $e_{t_1}, e_{t_2}, e_{t_3}, e_{t_4}, f_{t_5}$ involve respectively $\overline{2\ 4}, \overline{3\ 4}, \overline{2\ 5}, \overline{3\ 5}, \overline{4\ 5}$ with coefficient 1, and no other standard tabloids.

Consider the following vector

$v =$ 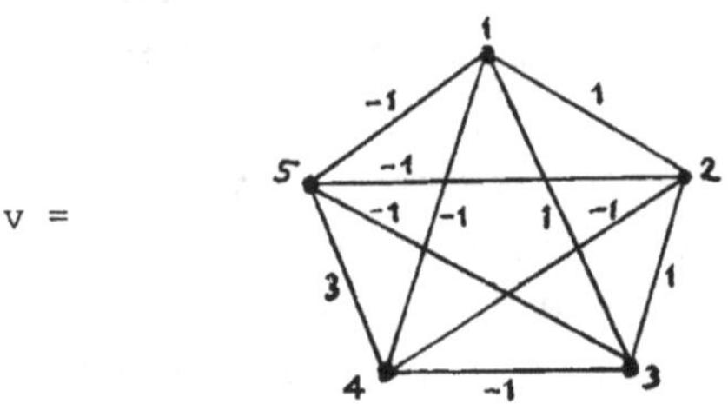

v belongs to $S^{(3,2)}$, since the sum of the edge coefficients is zero, and each point has valency zero (cf. 5.3). But v involves $-\overline{2\ 4}$, $-\overline{3\ 4}$, $-\overline{2\ 5}$, $-\overline{3\ 5}$, $3.\overline{4\ 5}$. Therefore

$$v = -e_{t_1} - e_{t_2} - e_{t_3} - e_{t_4} + 3f_{t_5}$$
$$= 2e_{t_1} - e_{t_2} - e_{t_3} - e_{t_4} + 3e_{t_5}.$$

Next we want the rather technical

8.14 LEMMA <u>Suppose that $\theta \epsilon \mathrm{Hom}_{\mathbb{Q}\mathfrak{S}_n}(M_{\mathbb{Q}}^{\lambda}, M_{\mathbb{Q}}^{\mu})$ and that all the tabloids involved in $\{t\}\theta$ have integer coefficients ($\{t\} \in M_{\mathbb{Q}}^{\lambda}$). Then, reducing all these integers modulo p, we obtain an element $\bar{\theta}$ of $\mathrm{Hom}_{F\mathfrak{S}_n}(M_F^{\lambda}, M_F^{\mu})$, where F is the field of p elements. If $\ker \theta = S_{\mathbb{Q}}^{\lambda\perp}$ then $\mathrm{Ker}\ \bar{\theta} \supseteq S_F^{\lambda\perp}$.</u>

<u>Proof</u>: It is trivial that $\bar{\theta} \in \mathrm{Hom}_{F\mathfrak{S}_n}(M_F^{\lambda}, M_F^{\mu})$.

Take a basis $f_1,\ldots,f_k$ of $S_{\mathbb{Q}}^{\lambda\perp}$ and extend by the standard basis of $S_{\mathbb{Q}}$ to obtain a basis $f_1,\ldots,f_m$ of $M_{\mathbb{Q}}^{\lambda}$. Let $\{t_1\},\ldots,\{t_m\}$ be the different λ-tabloids. Define the matrix $N = (n_{ij})$ by

$$n_{ij} = \langle f_i, \{t_j\} \rangle$$

We may assume that N has integer entries, and by row reducing the first k rows, we may assume that the first k rows of N (which correspond to the basis of $S_{\mathbb{Q}}^{\lambda\perp}$) are linearly independent modulo p. Reducing all the entries in N modulo p, we obtain a set of vectors in M_F^{λ}, the last m - k of which are the standard basis of S_F^{λ}, and the first k of which are linearly independent and orthogonal to the standard basis of S_F^{λ}. Since

$$\dim S_F^{\lambda\perp} = \dim M_F^{\lambda} - \dim S_F^{\lambda} = k ,$$

we have constructed a basis of $S_{\mathbb{Q}}^{\lambda\perp}$ whose elements give a basis of $S_F^{\lambda\perp}$ when the tabloid coefficients are reduced modulo p.

Now, any one of our basis elements of $S_{\mathbb{Q}}^{\lambda\perp}$ is an integral linear combination of λ-tabloids, and is sent to zero by θ. Therefore, when all integers are reduced modulo p, $\bar{\theta}$ certainly sends the basis of $S_F^{\lambda\perp}$ to zero, as required.

We can now complement Theorem 6.7 by proving

8.15 THEOREM <u>Over any field, $S^\lambda \otimes S^{(1^n)}$ is isomorphic to the dual of $S^{\lambda'}$.</u>

Proof: It is sufficient to consider the case where the ground field is F, the field of p elements, since we have proved the result when $F = \mathbb{Q}$.

In the proof of Theorem 6.7, we gave a $\mathbb{Q}\mathfrak{S}_n$-homomorphism Θ from $M_{\mathbb{Q}}^{\lambda'}$ into $M_{\mathbb{Q}}^{\lambda} \otimes S_{\mathbb{Q}}^{(1^n)}$ and proved that $\operatorname{Ker}\Theta = S_{\mathbb{Q}}^{\lambda'\perp}$. Using the Lemma above, $\bar{\Theta}$, defined by

$$\bar{\Theta}: \{t'\pi\} \to (\operatorname{sgn}\pi)\ \{t\pi\}\kappa_t\pi \otimes u$$

is an $F\mathfrak{S}_n$-homomorphism onto $S_F \otimes S_F^{(1^n)}$ whose kernel contains $S_F^{\lambda'\perp}$. By dimensions, $\operatorname{Ker}\bar{\Theta} = S_F^{\lambda'\perp}$, and the result follows.

9. THE BRANCHING THEOREM

The Branching Theorem tells us how to restrict an ordinary irreducible representation from $\mathfrak{S}_n$ to $\mathfrak{S}_{n-1}$. We have introduced the symbols $\downarrow \mathfrak{S}_{n-1}$ for restriction to $\mathfrak{S}_{n-1}$ and $\uparrow \mathfrak{S}_{n+1}$ for inducing to $\mathfrak{S}_{n+1}$. Using notation like that in 6.4, we have

9.1 EXAMPLE $[4,2^2,1]\downarrow \mathfrak{S}_8 = [3,2^2,1] + [4,2,1^2] + [4,2^2]$

$[4,2^2,1]\uparrow \mathfrak{S}_{10} = [5,2^2,1] + [4,3,2,1] + [4,2^3] + [4,2^2,1^2]$

These are special cases of

9.2 THE BRANCHING THEOREM

(i) $S^\mu_{\mathbb{Q}} \uparrow \mathfrak{S}_{n+1} \cong \oplus \{S^\lambda_{\mathbb{Q}} \mid [\lambda]$ is a diagram obtained by adding a node to $[\mu]$ $\}$.

(ii) $S^\mu_{\mathbb{Q}} \downarrow \mathfrak{S}_{n-1} \cong \oplus \{S^\lambda_{\mathbb{Q}} \mid [\lambda]$ is a diagram obtained by taking a node away from $[\mu]\}$.

Proof: The two parts of the Theorem are equivalent, by the Frobenius Reciprocity Theorem. Part (ii) follows from the more general:

9.3 THEOREM *When S^μ is defined over an arbitrary field, $S^\mu \downarrow \mathfrak{S}_{n-1}$ has a series with each factor isomorphic to a Specht module for $\mathfrak{S}_{n-1}$. The factors occurring are those given by part (ii) of the Branching Theorem, and S^{λ^i} occurs above S^{λ^j} in the series if $\lambda^i \rhd \lambda^j$.*

Proof: (See Peel [19]) Let $r_1 < r_2 < \ldots < r_m$ be the integers such that a node can be removed from the r_ith row of $[\mu]$ to leave a diagram (e.g. when $[\mu] = [4,2^2,1]$, $r_1, r_2, r_3 = 1, 3, 4$). Suppose that $[\lambda^i]$ is the diagram obtained by removing a node from the end of the r_ith row of $[\mu]$.

Define $\theta_i \in \mathrm{Hom}_{F\mathfrak{S}_{n-1}}(M^\mu, M^{\lambda^i})$ by

$$\theta_i : \{t\} \to \begin{cases} 0 & \text{if } n \notin r_i\text{th row of } \{t\} \\ \{\bar{t}\} & \text{if } n \in r_i\text{th row of } \{t\} \end{cases}$$

where $\{\bar{t}\}$ is $\{t\}$, with n removed.

When t is standard,

9.4 $$\theta_i : e_t \to \begin{cases} e_{\bar{t}} & \text{if } n \in r_i\text{th row of } t \\ 0 & \text{if } n \in r_1\text{th}, r_2\text{th}, \ldots, \text{or } r_{i-1}\text{th row of } t. \end{cases}$$

Let V_i be the space spanned by those polytabloids e_t where t is a standard μ-tableau and n is in the r_1th, r_2th, ..., or r_ith row of t. Then $V_{i-1} \subseteq \mathrm{Ker}\,\theta_i$ and $V_i\theta_i = S^{\lambda^i}$,

since the standard λ^i-polytabloids span S^{λ^i}.

In the series

$$0 \subseteq V_1 \cap \mathrm{Ker}\,\theta_1 \subset V_1 \subseteq V_2 \cap \mathrm{Ker}\,\theta_2 \subset V_2 \subseteq \ldots$$

$$\ldots \subset V_{m-1} \subseteq V_m \cap \text{Ker } \theta_m \subset V_m = S^{\mu}$$

we have $\dim(V_i/(V_i \cap \text{Ker } \theta_i)) = \dim V_i \theta_i = \dim S^{\lambda^i}$.

But

$$\sum_{i=1}^{m} \dim S^{\lambda^i} = \dim S^{\mu},$$

since the dimension of a Specht module is the number of standard tableaux. Therefore, there is equality in all possible places in the series above, and V_i/V_{i-1} is $F\mathfrak{S}_{n-1}$ - isomorphic to S^{λ^i}. This is our desired result.

9.5 EXAMPLE As an $F\mathfrak{S}_8$-module, $S^{(4,2^2,1)}$ has a series with factors, reading from the top, isomorphic to $S^{(4,2^2)}$, $S^{(4,2,1^2)}$, $S^{(3,2^2,1)}$. (cf. Example 17.16.)

10. p-REGULAR PARTITIONS

We have seen that $S^\mu/(S^\mu \cap S^{\mu\perp})$ is zero or irreducible, and that it can be zero only if the ground field has prime characteristic p. In order to distinguish between those partitions for which S^μ is or is not contained in $S^{\mu\perp}$, we make the following

10.1 DEFINITION A partition μ is <u>p-singular</u> if for some i

$$\mu_{i+1} = \mu_{i+2} = \ldots = \mu_{i+p} > 0.$$

Otherwise, μ is <u>p-regular</u>.

For example, $(6^2,5^4,1)$ is p-regular if and only if $p \geq 5$.

A conjugacy class of a group is called a <u>p-regular class</u> if the order of an element in that class is coprime to p.

10.2 LEMMA <u>The number of p-regular classes of $\mathfrak{S}_n$ equals the number of p-regular partitions of n.</u>

<u>Proof</u>: Writing a permutation π as a product of disjoint cycles, we see that π has order coprime to p if and only if no cycle has length divisible by p. Therefore, the number of p-regular classes of $\mathfrak{S}_n$ equals the number of partitions μ of n where no part μ_i of μ is divisible by p.

Now simplify the following ratio in two ways:

$$\frac{(1 - x^p)(1 - x^{2p})\ldots}{(1 - x)(1 - x^2)\ldots}$$

(i) Cancel equal factors $(1 - x^{mp})$ in the numerator and denominator. This leaves

$$\prod_{p\nmid i} (1 - x^i)^{-1} = \prod_{p\nmid i} (1 + x^i + (x^i)^2 + (x^i)^3 + \ldots)$$

and the coefficient of x^n is the number of partitions of n where no summand is divisible by p. (The partition $(\ldots 3^c, 2^b, 1^a)$ corresponds to taking x^a from the first bracket $(x^2)^b$ from the second bracket, and so on.)

(ii) For each m divide $(1 - x^m)$ in the denominator into $(1 - x^{mp})$ in the numerator, to give

$$\prod_{m=1}^{\infty} (1 + x^m + (x^m)^2 + \ldots + (x^m)^{p-1}).$$

Here the coefficient of x^n is the number of partitions of n where no part of the partition occurs p or more times.

Comparing coefficients of x^n, we obtain the desired equality (The reader who is worried about problems of convergence is referred to section 19.3 of Hardy and Wright [3]).

<u>Remark</u> Like most combinatorial results involving p-regularity, Lemma

10.2 does not require p to be prime, and it is only when we come to representation theory that we must not allow p to be composite.

We next want to investigate the integer g^μ defined by

<u>10.3</u> $g^\mu = \text{g.c.d.}\{\langle e_t, e_{t*}\rangle \mid e_t$ and e_{t*} are polytabloids in $S^\mu_{\mathbb{Q}}\}$.

The importance of this number is that it is the greatest common divisor of the entries in the Gram matrix with respect to the standard basis of the Specht module. (Corollary 8.6 shows that any polytabloid can be written as an <u>integral</u> linear combination of standard polytabloids).

10.4 LEMMA (James [7]) <u>Suppose that the partition μ has z_j parts equal to j. Then $\prod_{j=1}^{\infty} z_j!$ divides g^μ and g^μ divides $\prod_{j=1}^{\infty} (z_j!)^j$.</u>

<u>Remarks</u> Since $0! = 1$, there is no problem about taking infinite products. Some of the integers involved in the definition of g^μ may be zero or negative, but we adopt the convention that, for example, g.c.d. $\{-3,0,6\} = 3$.

<u>Proof</u>: Define an equivalence relation $\sim$ on the set of μ-tabloids by

$\{t_1\} \sim \{t_2\}$ if and only if for all i and j, i and j belong to the same row of $\{t_2\}$ when i and j belong to the same row of $\{t_1\}$.

Informally, this is saying that we can go from $\{t_1\}$ to $\{t_2\}$ by shuffling rows. The equivalence classes have size $\prod_{j=1}^{\infty} z_j!$

Now, if $\{t_1\}$ is involved in e_t and $\{t_1\} \sim \{t_2\}$, then the definition of a polytabloid shows that $\{t_2\}$ is involved in e_t, and whether the coefficients (which are ± 1) are the same or have opposite signs depends only on $\{t_1\}$ and $\{t_2\}$. Therefore, any two polytabloids have a multiple of $\prod_{j=1}^{\infty} z_j!$ tabloids in common, and $\prod_{j=1}^{\infty} z_j!$ divides g^μ (cf. Example 5.4).

Next, let t be any μ-tableau, and obtain t^* from t by reversing the order of the numbers in each row of t. For example,

$$\text{if } t = \begin{matrix} 1 & 2 & 3 & 4 \\ 5 & 6 & 7 & \\ 8 & 9 & 10 & \\ 11 & & & \end{matrix} \qquad \text{then } t^* = \begin{matrix} 4 & 3 & 2 & 1 \\ 7 & 6 & 5 & \\ 10 & 9 & 8 & \\ 11 & & & \end{matrix}$$

Let π be an element of the column stabilizer of t having the property that for every i, the numbers i and $i\pi$ belong to rows of t which have the same length. (In the example, π can be any element of the group $\mathfrak{S}_{\{5,8\}} \times \mathfrak{S}_{\{6,9\}} \times \mathfrak{S}_{\{7,10\}}$). Then $\{t\pi\}$ is involved in e_t and e_{t*} with the same coefficient in each. It is easy to see that all tabloids common to e_t and e_{t*} have this form. (In the example, every tabloid involved in e_{t*} has 1 in the first row. Looking at e_t, no common tabloid has 5 or 8 in the first row. Going back to e_{t*}, 2 must be in the first row of a common tabloid, and so on.) Therefore, $< e_t, e_{t*} > = \prod_{j=1}^{\infty} (z_j!)^j$, and the lemma is proved.

10.5 COROLLARY <u>The prime p divides g^μ if and only if μ is p-singular.</u>

<u>Proof</u>: μ is p-singular if and only if p divides $z_j!$ for some j, and this happens if and only if p divides g^μ.

10.6 COROLLARY <u>If t^* is obtained the μ-tableau t by reversing the order of the numbers in each row of t, then $e_{t*}\kappa_t$ is a multiple of e_t, and this multiple is coprime to p if and only if μ is p-regular.</u>

<u>Proof</u>: Corollary 4.7 shows that $e_{t*}\kappa_t$ is a multiple of e_t, $e_{t*}\kappa_t = h\, e_t$ say. Now,

$$h = h < e_t, \{t\} > = < h\, e_t, \{t\} > = < e_{t*}\kappa_t, \{t\} >$$
$$= < e_t{}^*, \{t\}\kappa_t > = < e_t{}^*, e_t > .$$

The last line of the proof of Lemma 10.4 shows that $h = \prod_{j=1}^{\infty}(z_j!)^j$, which is coprime to p if and only if μ is p-regular.

11. THE IRREDUCIBLE REPRESENTATIONS OF $\mathfrak{S}_n$

The ordinary irreducible representations of $\mathfrak{S}_n$ were constructed at the end of section 4. We now assume that our ground field has characteristic p, and the characteristic 0 case can be subsumed in this one, by allowing $p = \infty$.

11.1 THEOREM <u>Suppose that S^μ is defined over a field of characteristic p. Then $S^\mu/(S^\mu \cap S^{\mu\perp})$ is non-zero if and only if μ is p-regular.</u>

<u>Proof:</u> $S^\mu \subseteq S^{\mu\perp}$ if and only if $< e_t, e_{t^*} > = 0$ for every pair of polytabloids e_t and e_{t^*} in S^μ. But this is equivalent to p dividing the integer g^μ defined in 10.3, and Corollary 10.5 gives the desired result.

Shortly, we shall prove that all the irreducible $F\mathfrak{S}_n$-modules are given by the modules D^μ_F where

11.2 DEFINITION Suppose that the characteristic of F is p (prime or $= \infty$) and that μ is p-regular. Let $D^\mu_F = S^\mu_F/(S^\mu_F \cap S^{\mu\perp}_F)$.

As usual, we shall drop the suffix F when our results are independent of the field.

To prove that no two D^μ's are isomorphic, we need a generalization of Lemma 4.10, which said that S^λ is sent to zero by every element of $\mathrm{Hom}_{F\mathfrak{S}_n}(M^\lambda, M^\mu)$ unless $\lambda \trianglerighteq \mu$.

11.3 LEMMA <u>Suppose that λ and μ are partitions of n, and λ is p-regular. Let U be a submodule of M^μ and suppose that θ is a non-zero $F\mathfrak{S}_n$-homomorphism from S^λ into M^μ/U. Then $\lambda \trianglerighteq \mu$ and if $\lambda = \mu$, then $\mathrm{Im}\,\theta \subseteq (S^\mu + U)/U$.</u>

<u>Remark</u> The submodule U is insignificant in the proof of this result. The essential part of the Lemma says that, for λ p-regular, S^λ is sent to zero by every element of $\mathrm{Hom}_{F\mathfrak{S}_n}(S^\lambda, M^\mu)$ unless $\lambda \trianglerighteq \mu$. (cf. Corollary 13.17).

<u>Proof:</u> (See Peel [20]). Let t be a λ-tableau and reverse the order of the row entries in t to obtain the tableau t^*. By Corollary 10.6,

$$e_{t^*}\kappa_t = h\, e_t \text{ where } h \neq 0.$$

But $h\, e_t\theta = e_{t^*}\kappa_t\theta = e_{t^*}\theta\kappa_t$

Since $h \neq 0$ and θ is non-zero, $e_{t^*}\theta\kappa_t \neq U$. By Lemma 4.6, $\lambda \trianglerighteq \mu$, and if $\lambda = \mu$, then

$$e_t\theta = h^{-1}e_{t^*}\theta\kappa_t = \text{a multiple of } e_t + U \in (S^\mu + U)/U.$$

The result follows, because S^λ is generated by e_t.

11.4 COROLLARY <u>Suppose that λ and μ are partitions of n, and λ is p-regular. Let U be a submodule of M^μ and suppose that θ is a non-zero $F\mathfrak{S}_n$ homomorphism from D^λ into M^μ/U. Then $\lambda \trianglerighteq \mu$ and $\lambda \triangleright \mu$ if $U \supseteq S^\mu$.</u>

<u>Proof</u>: We can lift θ to a non-zero element of $\mathrm{Hom}_{F\mathfrak{S}_n}(S^\lambda, M^\mu/U)$ as follows:

$$S^\lambda \underset{\text{canon.}}{\rightarrow} S^\lambda/(S^\lambda \cap S^{\lambda\perp}) = D^\lambda \underset{\theta}{\rightarrow} M^\mu/U$$

Therefore, $\lambda \trianglerighteq \mu$, by the Lemma. If $\lambda = \mu$ then $\mathrm{Im}\,\theta$ is a non-zero submodule of $(S^\mu + U)/U$, so U does not contain S^μ.

11.5 THEOREM (James [7]) <u>Suppose that our ground field F has characteristic p (prime or $= \infty$). As μ varies over p-regular partitions of n, D^μ varies over a complete set of inequivalent irreducible $F\mathfrak{S}_n$-modules. Each D^μ is self-dual and absolutely irreducible. Every field is splitting field for $\mathfrak{S}_n$.</u>

<u>Proof</u>: Theorems 4.9 and 11.1 show that D^μ is self-dual and absolutely irreducible.

Suppose that $D^\lambda \cong D^\mu$. Then we have a non-zero $F\mathfrak{S}_n$-homomorphism from D^λ into $M^\lambda/(S^\mu \cap S^{\mu\perp})$, and by Corollary 11.4, $\lambda \trianglerighteq \mu$. Similarly, $\mu \trianglerighteq \lambda$, so $\lambda = \mu$.

Having shown that no two D^μ's are isomorphic, we are left with the question: Why have we got all the irreducible representations over F? In section 17 we shall prove that every composition factor of the regular representation over F is isomorphic to some D^μ, and then Theorem 1.1 gives our result. Rather than follow this artificial approach, the reader will probably prefer to accept two results from representation theory which we quote from Curtis and Reiner [2]:

<u>Curtis and Reiner 83.7:</u> If $\mathbb{Q}$ is a splitting field for a group G, then every field is a splitting field for G.

<u>Curtis and Reiner 83.5:</u> If F is a splitting field for G, then the number of inequivalent irreducible FG-modules equals the number of p-regular classes of G.

Since Theorem 4.12 shows $\mathbb{Q}$ is a splitting field, Lemma 10.2 now sees us home. More subtle, (to make use of our knowledge that D^μ is absolutely irreducible), is to combine Curtis and Reiner 83.5 with <u>Curtis and Reiner 82.6:</u> The number of inequivalent absolutely irreducible FG-modules is less than or equal to the number of p-regular classes of G.

Theorem 1.6 gives

11.6 THEOREM <u>The dimension of the irreducible representation D^μ of $\mathfrak{S}_n$ over a field of characteristic p can be calculated by evaluating the p-rank of the Gram matrix with respect to the standard basis of S^μ.</u>

11.7 EXAMPLE We have already illustrated an application of Theorem 11.6 in Example 5.2. Consider now the partition (2,2). The Gram matrix we obtain is (cf. Example 5.4):

$$A = \begin{bmatrix} 4 & 2 \\ 2 & 4 \end{bmatrix}$$

The p-rank of this is 0,1 or 2 if p = 2, 3 or >3, respectively. Therefore, $S^{(2,2)}/(S^{(2,2)} \cap S^{(2,2)\perp}) = 0$ if char F = 2, and dim $D^{(2,2)}$ = 1 or 2 if char F = 3 or >3, respectively.

11.8 THEOREM <u>The dimension of every non-trivial 2-modular irreducible representation of $\mathfrak{S}_n$ is even.</u>

<u>Proof</u>: If $\mu \neq (n)$ and t is a μ-tableau, then $< e_t, e_t >$, being the order of the column stabilizer of t, is even. Hence < , > is an alternating bilinear form when char F = 2, and it is well-known that an alternating bilinear form has even rank, so Theorem 11.6 gives the result.

<u>Remark</u> Theorem 11.8 is a special case of a general result which states that every non-trivial, self-dual, absolutely irreducible 2-modular representation of a group has even dimension.

The homomorphism $\bar{\theta}$ in the proof of Theorem 8.15 sends $\{t'\}\kappa_{t'}$ to $\{t\}\kappa_t\rho_t \otimes u$, and Ker $\bar{\theta} = S^{\lambda'\perp}$. Thus, if λ' is p-regular, the submodule of S^λ generated by $\{t\}\kappa_t\rho_t$ is isomorphic to $D^{\lambda'}$. In terms of the group algebra $F\mathfrak{S}_n$, this means that the right ideals generated by $\rho_t\kappa_t\rho_t$ (choosing one t for each partition whose conjugate is p-regular) give all the irreducible representations of $\mathfrak{S}_n$ over F when char F = p (p prime or = ∞).

12 COMPOSITION FACTORS

We next examine what can be said about the composition factors of M^μ and S^μ in general terms. When the ground field has characteristic zero, all the composition factors of M^μ are known (see section 14). The problem of finding the composition factors of S^μ when the field is of prime characteristic is still open. (All published algorithms for calculating the complete decomposition matrices for arbitrary symmetric groups give <u>incorrect</u> answers.)

First, a generalisation of Theorem 4.13:

12.1 THEOREM <u>All the composition factors of M^μ have the form D^λ with $\lambda \trianglerighteq \mu$, except if μ is p-regular, when D^μ occurs precisely once</u>.

<u>Proof</u>: Consider the following picture:

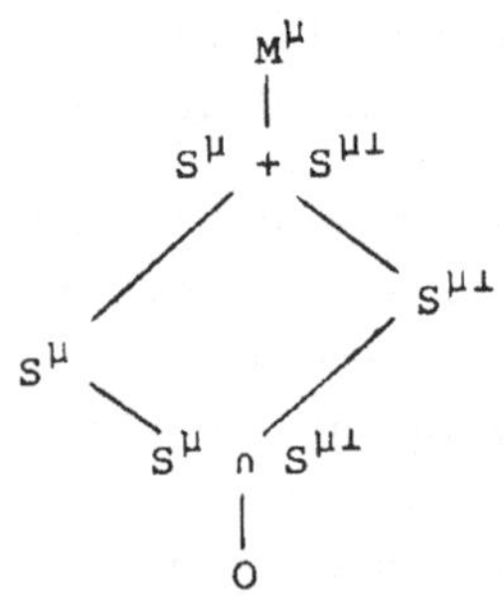

By Corollary 11.4, all the composition factors of M^μ/S^μ have the form D^λ with $\lambda \triangleright \mu$. But $S^{\mu\perp}$ is isomorphic to the dual of M^μ/S^μ, and so has the same composition factors, in the opposite order. (See 1.4, and recall that every irreducible $F\mathfrak{S}_n$-module is self-dual.) Now, $S^\mu/(S^\mu \cap S^{\mu\perp})$ is non-zero if and only if μ is p-regular, when it equals D^μ. Since $0 \subseteq S^\mu \cap S^{\mu\perp} \subseteq S^\mu \subseteq M^\mu$ is a series for M^μ, the Theorem is proved.

12.2 COROLLARY <u>If μ is p-regular, S^μ has a unique top composition factor $D^\mu = S^\mu/(S^\mu \cap S^{\mu\perp})$. If D is a composition factor of $S^\mu \cap S^{\mu\perp}$ then $D \cong D^\lambda$ for some $\lambda \triangleright \mu$. If μ is p-singular, all the composition factors of S^μ have the form D^λ with $\lambda \triangleright \mu$.</u>

<u>Proof</u>: This is an immediate corollary of Theorems 4.9 and 12.1.

The decomposition matrix of a group records the multiplicities of the p-modular irreducible representations in the reductions modulo p of the ordinary irreducible representations. Corollaries 8.11 and 12.2 give

12.3 COROLLARY <u>The decomposition matrix of $\mathfrak{S}_n$ for the prime p has the form:</u>

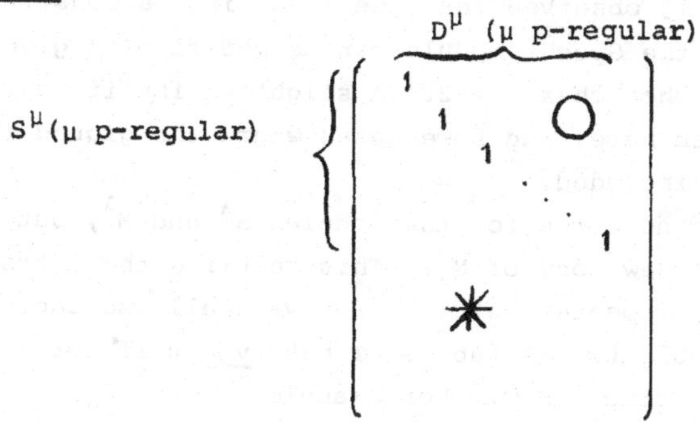

<u>when the p-regular partitions are placed in dictionary order before all the p-singular partitions.</u>

12.4 EXAMPLE Consider n = 3, $S^{(3)} = D^{(3)}$ is the trivial p-modular representation. $S^{(1^3)}$ is the alternating representation, and $S^{(1^3)} \cong S^{(3)}$ if and only if p = 2. Using Example 5.1, the decomposition matrices of $\mathfrak{S}_3$ are:

$$\begin{array}{c} \\ S^{(3)} \\ S^{(2,1)} \\ S^{(1^3)} \end{array} \begin{array}{c} \begin{array}{cc} D^{(3)} & D^{(2,1)} \end{array} \\ \left(\begin{array}{cc} 1 & \\ & 1 \\ 1 & \end{array}\right) \end{array} \text{ when } p = 2, \qquad \begin{array}{c} \\ S^{(3)} \\ S^{(2,1)} \\ S^{(1^3)} \end{array} \begin{array}{c} \begin{array}{cc} D^{(3)} & D^{(2,1)} \end{array} \\ \left(\begin{array}{cc} 1 & \\ 1 & 1 \\ & 1 \end{array}\right) \end{array} \text{ when } p = 3$$

$$\begin{array}{c} \\ S^{(3)} \\ S^{(2,1)} \\ S^{(1^3)} \end{array} \begin{array}{c} \begin{array}{ccc} D^{(3)} & D^{(2,1)} & D^{(1^3)} \end{array} \\ \left(\begin{array}{ccc} 1 & & \\ & 1 & \\ & & 1 \end{array}\right) \end{array} \text{ when } p > 3$$

(By convention, omitted matrix entries are always zero.)

13 SEMISTANDARD HOMOMORPHISMS

Carter and Lusztig [1] observed that the ideas in the construction of the standard basis of the Specht module can be modified to give a basis for $\mathrm{Hom}_{F\mathfrak{S}_n}(S^\lambda, M^\mu)$ when char $F \neq 2$. A slightly simplified form of their argument is given here, and some cases where the ground field has characteristic 2 are included.

We keep our previous notation for the modules S^λ and M^λ, but it is convenient to introduce a new copy of M^μ. This requires the introduction of tableaux T having repeated entries, and we shall use capital letters to denote such tableaux. A tableau T has <u>type</u> μ if for every i, the number i occurs μ_i times in T. For example

$$\begin{matrix} 2 & 2 & 1 & 1 \\ 1 & & & \end{matrix}$$

is a (4,1)-tableau of type (3,2).

13.1 DEFINITION $\mathcal{T}(\lambda,\mu) = \{T \mid T \text{ is a } \lambda\text{-tableau of type } \mu\}$.

<u>Remark</u>: We allow μ to be any sequence of non-negative integers, whose sum is n. For example, if n = 10, μ can be (4,5,0,1). The definition of M^μ as the permutation module of $\mathfrak{S}_n$ on a Young subgroup does not require $\mu_1 \geq \mu_2 \geq \ldots$, and $M^{(4,5,0,1)} \cong M^{(5,4,1)}$.

<u>For the remainder of section 13, let t be a given λ-tableau (of type (1^n))</u>.

If $T \in \mathcal{T}(\lambda,\mu)$, let (i)T be the entry in T which occurs in the same position as i occurs in t. Let $\mathfrak{S}_n$ act on $\mathcal{T}(\lambda,\mu)$ by

$$(i)(T\pi) = (i\pi^{-1})T \qquad (1 \leq i \leq n,\ T \in \mathcal{T}(\lambda,\mu), \pi \in \mathfrak{S}_n).$$

The action of π is therefore that of a place permutation, and we are forced to take π^{-1} in the definition to make the $\mathfrak{S}_n$-action well-defined.

13.2 EXAMPLE If $t = \begin{matrix} 1 & 3 & 4 & 5 \\ 2 & & & \end{matrix}$ and $T = \begin{matrix} 2 & 2 & 1 & 1 \\ 1 & & & \end{matrix}$ then

$$T(1\ 2) = \begin{matrix} 1 & 2 & 1 & 1 \\ 2 & & & \end{matrix} \quad \text{and} \quad T(1\ 2\ 3) = \begin{matrix} 2 & 1 & 1 & 1 \\ 2 & & & \end{matrix}.$$

Since $\mathfrak{S}_n$ is transitive on $\mathcal{T}(\lambda,\mu)$, and the stabilizer of an element is a Young subgroup $\mathfrak{S}_\mu$, we may take M^μ to be the vector space over F spanned by the tableaux in $\mathcal{T}(\lambda,\mu)$. It will soon emerge why we have defined M^μ in a way which depends on both λ and μ.

If T_1 and T_2 belong to $\mathcal{T}(\lambda,\mu)$, we say that T_1 and T_2 are <u>row</u> (respectively, <u>column</u>) <u>equivalent</u> if $T_2 = T_1\pi$ for some permutation π in the row (respectively, column) stabilizer of the given λ-tableau t.

13.3 DEFINITION If $T \in \mathcal{T}(\lambda,\mu)$, define the map θ_T by $\theta_T : \{t\}S \to \Sigma\{T_1 | T_1 \text{ is row equivalent to } T\}S \quad (S \in F\mathfrak{S}_n)$.

It is easy to verify that θ_T belongs to $\mathrm{Hom}_{F\mathfrak{S}_n}(M^\lambda, M^\mu)$.

13.4 EXAMPLE If $t = \begin{matrix} 1 & 3 & 4 & 5 \\ 2 \end{matrix}$ and $T = \begin{matrix} 2 & 2 & 1 & 1 \\ 1 \end{matrix}$ then

$$\{t\}\theta_T = \begin{matrix} 2 & 2 & 1 & 1 \\ 1 \end{matrix} + \begin{matrix} 2 & 1 & 2 & 1 \\ 1 \end{matrix} + \begin{matrix} 2 & 1 & 1 & 2 \\ 1 \end{matrix} + \begin{matrix} 1 & 2 & 2 & 1 \\ 1 \end{matrix} + \begin{matrix} 1 & 2 & 1 & 2 \\ 1 \end{matrix} + \begin{matrix} 1 & 1 & 2 & 2 \\ 1 \end{matrix} \quad \text{and}$$

$$\{t\}(123)\theta_T = \begin{matrix} 2 & 1 & 1 & 1 \\ 2 \end{matrix} + \begin{matrix} 1 & 1 & 2 & 1 \\ 2 \end{matrix} + \begin{matrix} 1 & 1 & 1 & 2 \\ 2 \end{matrix} + \begin{matrix} 2 & 1 & 2 & 1 \\ 1 \end{matrix} + \begin{matrix} 2 & 1 & 1 & 2 \\ 1 \end{matrix} + \begin{matrix} 1 & 1 & 2 & 2 \\ 1 \end{matrix}$$

Notice that the way to write down $\{t\}\theta_T$ is simply to sum all the different tableaux whose rows contain the same numbers as the corresponding row of T.

It is clear that

<u>13.5 $T\kappa_t = 0$ if and only if some column of T contains two identical numbers.</u>

If we define $\hat{\theta}_T$ by

$$\hat{\theta}_T = \text{the restriction of } \theta_T \text{ to } S^\lambda,$$

then 13.5 suggests that sometimes $\hat{\theta}_T$ is zero, since $e_t\hat{\theta}_T = \{t\}\hat{\theta}_T\kappa_t$. To eliminate such trivial elements of $\mathrm{Hom}_{F\mathfrak{S}_n}(S^\lambda, M^\mu)$, we make the following

13.6 DEFINITION A tableau T is <u>semistandard</u> if the numbers are non-decreasing along the rows of T and strictly increasing down the columns of T. Let $\mathcal{T}_0(\lambda,\mu)$ be the set of semistandard tableaux in $\mathcal{T}(\lambda,\mu)$.

13.7 EXAMPLE If $\lambda = (4,1)$ and $\mu = (2,2,1)$, then $\mathcal{T}_0(\lambda,\mu)$ consists of the two tableaux $\begin{matrix} 1 & 1 & 2 & 2 \\ 3 \end{matrix}$ and $\begin{matrix} 1 & 1 & 2 & 3 \\ 2 \end{matrix}$.

We aim to prove that the homomorphisms $\hat{\theta}_T$ with T in $\mathcal{T}_0(\lambda,\mu)$

usually give a basis for $\mathrm{Hom}_{F\mathfrak{S}_n}(S^\lambda, M^\mu)$. These homomorphisms will be called <u>semistandard homomorphisms</u>, and, as with the standard basis of the Specht module, the difficult part is to decide whether the semistandard homomorphisms <u>span</u> $\mathrm{Hom}_{F\mathfrak{S}_n}(S^\lambda, M^\mu)$. The proof that they are linearly independent uses a partial order on the column equivalence classes [T] of tableaux in $\mathcal{T}(\lambda,\mu)$ (cf. 3.11 and 3.15):

13.8 DEFINITION Let $[T_1] \lhd [T_2]$ if $[T_2]$ can be obtained from $[T_1]$ by interchanging w and x, where w belongs to a later column of T_1 than x and $w < x$. Then $\lhd$ generates a partial order $\lhd$.

13.9 EXAMPLE When $\lambda = (3,2)$ and $\mu = (2,2,1)$, the following tree indicates the partial order on the column equivalence classes:

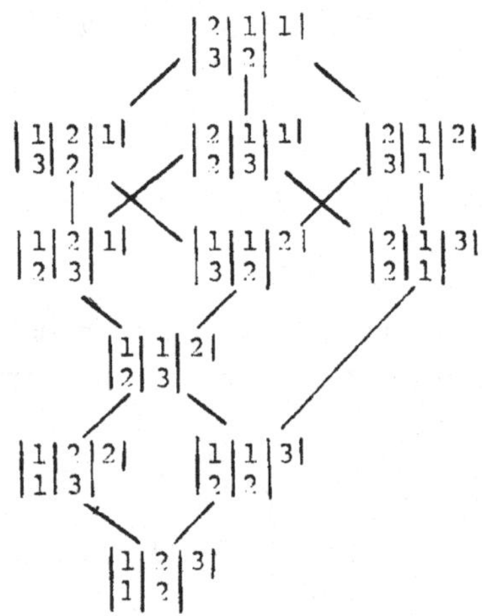

The crucial, but trivial, property of this partial order is:

<u>13.10</u> It T is semistandard, and T' is row equivalent to T, then $[T'] \lhd [T]$ unless $T' = T$.

13.11 LEMMA <u>$\{\hat{\Theta}_T \mid T \in \mathcal{T}_0(\lambda,\mu)\}$ is a linearly independent subset of $\mathrm{Hom}_{F\mathfrak{S}_n}(S^\lambda, M^\mu)$.</u>

<u>Proof</u>: (cf. Lemmas 8.2 and 8.3). If $\Sigma a_T \Theta_T$ is a linear combination of homomorphisms with T in $\mathcal{T}_0(\lambda,\mu)$ and not all the field coefficients equal zero, choose T_1 such that $a_{T_1} \neq 0$, but $a_T = 0$ if $[T_1] \lhd [T]$. Then from the definition of Θ_T and 13.10,

$$\{t\} \Sigma a_T \Theta_T = a_{T_1} T_1 + \text{a linear combination of tableaux } T_2 \text{ satisfying } [T_1] \not\ntrianglerighteq [T_2].$$

Since the column stabilizer of t preserves column equivalence classes, and $T_1 \kappa_t \neq 0$, this shows that

$$\{t\}\kappa_t \; \Sigma a_T \, \theta_T = \{t\} \; \Sigma a_T \, \theta_T \, \kappa_t \neq 0$$

Therefore, $\Sigma a_T \hat{\theta}_T$ is a non-zero element of $\mathrm{Hom}_{F\mathfrak{S}_n}(S^\lambda, M^\mu)$, as required.

We now have to be careful about the case where our ground field has characteristic 2:

13.12 LEMMA <u>Suppose that θ is a non-zero element of $\mathrm{Hom}_{F\mathfrak{S}_n}(S^\lambda, M^\mu)$, and write</u>

$$\{t\}\kappa_t \theta = \Sigma c_T \, T \qquad (c_T \in F, \; T \in \mathcal{T}(\lambda,\mu))$$

<u>where t is the given λ-tableau. Unless char F = 2 and λ is 2-singular, then</u>

(i) <u>$c_{T^*} = 0$ for every tableau T^* having a repeated entry in some column.</u>

<u>and (ii) $c_{T_1} \neq 0$ for some semistandard tableau T_1.</u>

<u>Proof</u>: <u>Part (i)</u> Suppose that $i \neq j$ are in the same column of t, and $(i)T^* = (j)T^*$. We wish to prove that $c_{T^*} = 0$.

Since $\kappa_t(i,j) = -\kappa_t$,

$$\Sigma \, c_T \, T(i,j) = \{t\}\kappa_t \, \theta(i,j) = -\Sigma \, c_T \, T$$

Because $T^*(i,j) = T^*$, it follows that $c_{T^*} = 0$ when char $F \neq 2$.

If char F = 2 and λ is 2-regular, let π be the permutation reversing the order of the numbers in each row of t. By Corollary 10.6 ,

$$\{t\}\kappa_t \, \pi \, \kappa_t = \{t\}\kappa_t \, .$$

Therefore

$$\Sigma \, c_T T = \{t\}\kappa_t \, \theta = \{t\}\kappa_t \theta \, \pi \, \kappa_t = \Sigma \, c_T \, T\pi \, \kappa_t \, .$$

By 13.5, no tableau which has a column containing a repeated entry appears in $\Sigma \, c_T \, T \, \pi \, \kappa_t$, so $c_{T^*} = 0$.

<u>Part (ii)</u> If π is in the column stabilizer of t, then $1 - (\mathrm{sgn}\,\pi)\pi$ annihilates $\{t\}\kappa_t$. Therefore

$$\Sigma \, c_T \, T = \Sigma \, c_T (\mathrm{sgn}\,\pi) T\pi \, ,$$

and so

$$c_{T_1} = \pm \, c_{T_2} \quad \text{when } T_1 \text{ and } T_2 \text{ are column equivalent.}$$

Since $\theta \neq 0$, we may choose a tableau T_1 such that $c_{T_1} \neq 0$, but $c_T = 0$ if $[T_1] \lhd [T]$. The previous paragraph and part (i) of the Lemma show that we may assume that the numbers strictly increase down the columns of T_1 .

We shall be home if we can derive a contradiction from assuming that for some j, $a_1 < a_2 < \ldots < a_r$ are the entries in the jth column of T_1, $b_1 < b_2 < \ldots < b_s$ are the entries in (j+1)th column of T_1 and $a_q > b_q$ for some q.

$$\begin{array}{ccc} a_1 & & b_1 \\ \cdot & & \wedge \\ \cdot & & \cdot \\ \cdot & & \wedge \\ a_q & > & b_q \\ \wedge & & \cdot \\ \cdot & & \cdot \\ \cdot & & \cdot \\ \cdot & & b_s \\ \wedge & & \\ a_r & & \end{array}$$

Let x_{ij} be the entry in the (i,j)th place of the tableau t, and let $\Sigma(\text{sgn }\sigma)\sigma$ be a Garnir element for the sets $\{x_{qj},\ldots,x_{rj}\}$ and $\{x_{1,j+1},\ldots,x_{q,j+1}\}$. Then

$$\Sigma\, c_T\, T\, \Sigma(\text{sgn }\sigma)\sigma = \{t\}\kappa_t\, \Sigma(\text{sgn }\sigma)\sigma\Theta = 0.$$

For every tableau T, $T\,\Sigma(\text{sgn }\sigma)\sigma$ is a linear combination of tableaux agreeing with T on all except the (1,j+1)th, (2,j+1)th,..., (q,j+1)th, (q,j)th,...,(r,j)th places. All the tableaux involved in $T_1\,\Sigma(\text{sgn }\sigma)\sigma$ have coefficient $\pm\, c_{T_1}$, and since $\Sigma\, c_T\, T\,\Sigma(\text{sgn }\sigma)\sigma$ is zero, there must be a tableau $T \neq T_1$ with $c_T \neq 0$ such that T agrees with T_1 on all except the places described above. Since $b_1 < \ldots < b_q < a_q < \ldots < a_r$, we must have $[T_1] \lhd [T]$, and this contradicts our initial choice of T_1.

13.13 THEOREM <u>Unless char F = 2 and λ is 2-singular,</u> <u>$\{\hat\theta_T \mid T \in \mathcal{T}_0(\lambda,\mu)\}$ is a basis for $\text{Hom}_{F\mathfrak{S}_n}(S^\lambda, M^\mu)$.</u>

<u>Proof:</u> Suppose Θ is a non-zero element of $\text{Hom}_{F\mathfrak{S}_n}(S^\lambda, M^\mu)$. By Lemma 13.12,

$$\{t\}\kappa_t\,\Theta = \Sigma c_T\, T, \text{ where } c_{T_1} \neq 0 \text{ for some } T_1 \in \mathcal{T}_0(\lambda,\mu).$$

We may assume that $c_T = 0$ if $T \in \mathcal{T}_0(\lambda,\mu)$ and $[T_1] \lhd [T]$. Then, by 13.10, $\{t\}\kappa_t(\Theta - c_{T_1}\hat\theta_{T_1})$ is a linear combination of tableaux T_2 with $[T_1] \ntrianglelefteq [T_2]$. By induction, $\Theta - c_{T_1}\hat\theta_{T_1}$ is a linear combination of semistandard homomorphisms, and so the same is true of Θ. The Theorem now follows from Lemma 13.11.

13.14 COROLLARY <u>Unless char F = 2 and λ is 2-singular,</u> <u>dim $\text{Hom}_{F\mathfrak{S}_n}(S^\lambda, M^\mu)$ equals the number of semistandard λ-tableaux of type μ</u> .

<u>Remark</u> If ν is obtained from μ by reordering the parts (e.g. μ =

(4,5,0,1) and $\nu = (5,4,1)$), then visibly

$$\dim \mathrm{Hom}_{F\mathfrak{S}_n}(S^\lambda, M^\mu) = \dim \mathrm{Hom}_{F\mathfrak{S}_n}(S^\lambda, M^\nu)$$

Equivalently, we may choose an unusual order of integers in definition 13.6. Therefore, the number of semistandard tableaux of a given shape and size is independent of the order we choose on the entries. For example, we list below the elements in $\mathcal{T}_0((4,1),(2,2,1))$ for different orderings of {1,2,3}:

1 1 2 2 3	1 1 2 3 2	when $1 < 2 < 3$
3 2 1 1 2	3 2 2 1 1	when $3 < 2 < 1$
1 1 3 2 2	1 1 2 2 3	when $1 < 3 < 2$

13.15 COROLLARY <u>Unless char F = 2 and λ is 2-singular, every element of $\mathrm{Hom}_{F\mathfrak{S}_n}(S^\lambda, M^\mu)$ can be extended to an element of $\mathrm{Hom}_{F\mathfrak{S}_n}(M^\lambda, M^\mu)$.</u>

<u>Proof</u>: $\hat{\Theta}_T$ can be extended to Θ_T.

Of course, Corollary 13.15 is trivial if char F = 0, but we know of no direct proof for the general case.

That Theorem 13.13 and Corollary 13.15 can be false if char F = 2 and λ is 2-singular is illustrated by the easy:

13.16 EXAMPLE If char F = 2, $\frac{\bar{1}}{2} + \frac{\bar{2}}{1} \to \overline{1\ 2}$ defines an element of $\mathrm{Hom}_{F\mathfrak{S}_2}(S^{(1^2)}, M^{(2)})$ which cannot be extended to an element of $\mathrm{Hom}_{F\mathfrak{S}_2}(M^{(1^2)}, M^{(2)})$.

13.17 COROLLARY <u>Unless char F = 2 and λ is 2-singular, $\lambda \not\trianglerighteq \mu$ implies $\mathrm{Hom}_{F\mathfrak{S}_n}(S^\lambda, M^\mu) = 0$, and $\mathrm{Hom}_{F\mathfrak{S}_n}(S^\lambda, M^\lambda) \cong F$.</u>

<u>Proof</u>: There is just one semistandard λ-tableau of type μ if $\lambda = \mu$, and none at all unless $\lambda \trianglerighteq \mu$. (cf. the proof of Lemma 3.7). Corollary 13.14 gives the result.

Corollary 13.17 has already been proved under the hypothesis that λ is p-regular (Lemma 11.3), and we now provide an alternative proof for the case where char F $\neq$ 2.

Let $\Theta \in \mathrm{Hom}_{F\mathfrak{S}_n}(S^\lambda, M^\mu)$, and suppose that t is a λ-tableau and t_1 is a μ-tableau. If $\lambda \not\trianglerighteq \mu$, or if $\lambda = \mu$ and $\{t_1\}$ is not involved in e_t, then some pair of numbers a,b belong to the same row of t_1 and the same column of t. Therefore

$$< e_t\theta,\{t_1\} > = -< e_t(a,b)\theta,\{t_1\} >$$
$$= -< e_t\theta,\{t_1\}(a,b) >$$
$$= -< e_t\theta,\{t_1\} > .$$

Since char F $\neq$ 2, $< e_t\theta,\{t_1\} > = 0$. This proves that $\theta = 0$ if $\lambda \ntrianglerighteq \mu$, and that $e_t\theta$ involves only tabloids involved in e_t when $\lambda = \mu$. If $\lambda = \mu$ and π belongs to the column stabilizer of t, then $< e_t\theta,\{t\}\pi > = < e_t\theta\ \pi^{-1},\{t\} > = \operatorname{sgn}\pi < e_t\theta,\{t\} >$ and this shows that $e_t\theta = < e_t\theta,\{t\} > e_t$. Thus θ is multiplication by a constant.

13.18 COROLLARY <u>Unless char F = 2 and λ is 2-singular, S^λ is indecomposable.</u>

<u>Proof</u>: If S^λ were decomposable, we could take the projection onto one component, and produce a non-trivial element of $\operatorname{Hom}_{F\mathfrak{S}_n}(S^\lambda,M^\lambda)$, contradicting the last Corollary.

<u>Remark</u>: There are decomposable Specht modules - see Example 23.10(iii).

When we investigate the representation theory of the general linear group, we shall need the simple

13.19 THEOREM <u>$\{\theta_T \mid T \in \mathcal{T}(\lambda,\mu)$ and the numbers are non-decreasing along each row of T$\}$ is a basis for $\operatorname{Hom}_{F\mathfrak{S}_n}(M^\lambda,M^\mu)$.</u>

<u>Proof</u>: Our set of homomorphisms has been constructed by taking one representative $T_1,T_2,\ldots,T_k$ from each row equivalence class of $\mathcal{T}(\lambda,\mu)$. The linear independence of the set follows from the definition of θ_T.

Suppose that θ is an element of $\operatorname{Hom}_{F\mathfrak{S}_n}(M^\lambda,M^\mu)$. If T and T' are row equivalent, then $T' = T\pi$ for some π in R_t, and so

$$< \{t\}\theta,T' > = < \{t\}\theta,T\pi > = < \{t\}\theta\pi^{-1},T >$$
$$= < \{t\}\theta,T >$$

Hence
$$\{t\}\theta = \sum_{i=1}^{k} < \{t\}\theta,T_i > \{t\}\theta_{T_i}$$

and since M^λ is a cyclic module, θ is a linear combination of θ_{T_i}'s as required :

$$\theta = \sum_{i=1}^{k} < \{t\}\theta,T_i > \theta_{T_i} .$$

14 YOUNG'S RULE

It is now possible to describe the composition factors of $M^{\mu}_{\mathbb{Q}}$ explicity.

14.1 YOUNG'S RULE <u>The multiplicity of $S^{\lambda}_{\mathbb{Q}}$ as a composition factor of $M^{\mu}_{\mathbb{Q}}$ equals the number of semistandard λ-tableaux of type μ.</u>

<u>Proof</u>: Since $\mathbb{Q}$ is a splitting field for $\mathfrak{S}_n$, the number we seek is $\dim \mathrm{Hom}_{\mathbb{Q}\mathfrak{S}_n}(S^{\lambda},M^{\mu})$, by 1.7. But this is equal to the number of semistandard λ-tableaux of type μ, by Corollary 13.14.

<u>Remark</u>: An independent proof of Young's Rule appears in section 17.

Young's Rule shows that the composition factors of $M^{\mu}_{\mathbb{Q}}$ are obtained by writing down all the semistandard tableaux of type μ which have the shape of a partition diagram.

14.2 EXAMPLE We calculate the factors of $M^{(3,2,2)}_{\mathbb{Q}}$. The semistandard tableaux of type μ are:

```
1 1 1 2 2 3 3        1 1 1 2 2 3        1 1 1 2 2
                     3                  3 3

1 1 1 2 3 3          1 1 1 2 3          1 1 1 2 3
2                    2 3                2
                                        3

1 1 1 2              1 1 1 2            1 1 1 3 3
2 3 3                2 3                2 2
                     3

1 1 1 3              1 1 1 3            1 1 1
2 2 3                2 2                2 2 3
                     3                  3

1 1 1
2 2
3 3
```

Therefore in the notation of 6.4,

$$[3][2][2] = [7] + 2[6,1] + 3[5,2] + 2[4,3] + [5,1^2] + 2[4,2,1] + [3^2,1] + [3,2^2]$$

<u>Remark</u>: Young's Rule gives the same answer whichever way we choose to

order the integers in the definition of "semistandard", and does not require μ to be a proper partition:

14.3 EXAMPLE The factors of $M_{\mathbb{Q}}^{(3,2)}$ are given by

by 1 1 1 2 2   1 1 1 2   1 1 1
                         2   2 2

or by 2 2 1 1 1   2 2 1 1   2 2 1
                           1   1 1

Therefore, [3][2] = [5] + [4,1] + [3,2] (cf. Example 5.2).

14.4 EXAMPLE If $m \leq n/2$ then

$$[n-m][m] = [n] + [n-1,1] + [n-2,2]+ \dots +[n-m,m].$$

Since $\dim M^{(n-m,m)} = \binom{n}{m}$, we deduce that

$$\dim S^{(n-m,m)} = \binom{n}{m} - \binom{n}{m-1}.$$

Notice that Young's Rule gives $S_{\mathbb{Q}}^{\mu}$ as a composition factor of $M_{\mathbb{Q}}^{\mu}$ with multiplicity one, and the other Specht modules $S_{\mathbb{Q}}^{\lambda}$ we get satisfy $\lambda \rhd \mu$ in agreement with Theorem 4.13. Remembering that this shows that the matrix $m = (m_{\lambda\mu})$ recording factors of $M_{\mathbb{Q}}^{\lambda}$ as λ varies (see 6.1) is lower triangular with 1's down the diagonal, we can use Young's Rule to write a given $[\mu]$ as a linear combination of terms of the form $[\lambda_1][\lambda_2]\dots[\lambda_j]$ (The method of doing this explicitly is given by the Determinantal Form - see section 19). Hence we can calculate terms like $[\mu][\nu_1]\dots[\nu_k]$ ($= S_{\mathbb{Q}}^{\mu} \otimes S_{\mathbb{Q}}^{(\nu_1)} \otimes \dots \otimes S_{\mathbb{Q}}^{(\nu_k)} \uparrow \mathfrak{S}_n$) for integers $\nu_1,\dots,\nu_k$. More generally, Young's Rule enables us to evaluate $[\mu][\nu]$($= S_{\mathbb{Q}}^{\mu} \otimes S_{\mathbb{Q}}^{\nu} \uparrow \mathfrak{S}_n$) for any pair of partitions μ and ν . The product $[\mu][\nu]$ is the subject of the Littlewood-Richardson Rule (section 16), and the argument we have just given shows that the Littlewood-Richardson Rule is a purely <u>combinatorial</u> generalisation of Young's Rule.

14.5 EXAMPLE We calculate $[3,2][2] = S_{\mathbb{Q}}^{(3,2)} \otimes S_{\mathbb{Q}}^{(2)} \uparrow \mathfrak{S}_7$ using only Young's Rule. By Example 14.4,

$$[3,2] = [3][2] - [4][1] .$$

To find [4][1][2], we use Young's Rule:

1 1 1 1 2 3 3   1 1 1 1 2 3   1 1 1 1 2
                             3   3 3

```
1 1 1 1 3 3      1 1 1 1 3      1 1 1 1 3
2                2 3            2
                                3

1 1 1 1          1 1 1 1
2 3 3            2 3
                 3
```

$[3,2][2] = [3][2][2] - [4][1][2]$, and using Example 14.2, we have
$[3,2][2] = [7] + 2[6,1] + 3[5,2] + 2[4,3] + [5,1^2] + 2[4,2,1]$
$+ [3^2,1] + [3,2^2] - [7] - 2[6,1] - 2[5,2] - [4,3] - [5,1^2] - [4,2,1]$
$= [5,2] + [4,3] + [4,2,1] + [3^2,1] + [3,2^2]$. (cf. Example 16.6).

15 SEQUENCES

In order to state the Littlewood-Richardson Rule in the next section, we must discuss properties of finite sequences of integers. A sequence is said to have <u>type</u> μ if, for each i, i occurs μ_i times in the sequence.

15.1 EXAMPLE The sequences of type (3,2) are

2 2 1 1 1	2 1 2 1 1	2 1 1 2 1	2 1 1 1 2	1 2 2 1 1
× × ✓ ✓ ✓	× ✓ ✓ ✓ ✓	× ✓ ✓ ✓ ✓	× ✓ ✓ ✓ ✓	✓ ✓ × ✓ ✓
1 2 1 2 1	1 2 1 1 2	1 1 2 2 1	1 1 2 1 2	1 1 1 2 2
✓ ✓ ✓ ✓ ✓	✓ ✓ ✓ ✓ ✓	✓ ✓ ✓ ✓ ✓	✓ ✓ ✓ ✓ ✓	✓ ✓ ✓ ✓ ✓

15.2 DEFINITION Given a sequence, the quality of each term is determined as follows (each term in a sequence is either <u>good</u> or <u>bad</u>).

(i) All the 1's are good.

(ii) An i + 1 is good if and only if the number of previous good i's is strictly greater than the number of previous good (i+1)'s.

15.3 EXAMPLES We have indicated the quality of the terms in the sequences of type (3,2) above. Here is another example:

3 1 1 2 3 3 2 3 2 1 2
× ✓ ✓ ✓ ✓ × ✓ ✓ × ✓ ✓

It follows immediately from the definition that an i+1 is bad if and only if the number of previous good i's equals the number of previous good (i+1)'s. Hence we have a result which will be needed later:

<u>15.4</u> If a sequence contains m good (c-1)'s in succession, then the next m c's in the sequence are all good.

15.5 DEFINITION Let $\mu = (\mu_1, \mu_2, \ldots)$ be a sequence of non-negative integers whose sum is n, and let $\mu^{\#} = (\mu_1^{\#}, \mu_2^{\#}, \ldots)$ be a sequence of non-negative integers such that for all i,

$$\mu_{i+1}^{\#} \le \mu_i^{\#} \le \mu_i .$$

Then $\mu^{\#}$, μ is called a <u>pair of partitions for n.</u>

<u>Remark:</u> As here, we shall frequently drop the condition $\mu_1 \ge \mu_2 \ge \ldots$ on a partition μ, but will still refer to μ as a partition of n. If the condition $\mu_1 \ge \mu_2 \ge \ldots$ holds we shall call μ a <u>proper partition</u> of n. So, for example , $\mu^{\#}$ is a proper partition of some $n' \le n$ in definition 15.5. Note that a Specht module S^{μ} is defined only for μ

a proper partition, but the module M^{μ} spanned by μ-tabloids may have μ improper.

15.6 DEFINITION Given a pair of partitions $\mu^{\#}$, μ for n, let $s(\mu^{\#},\mu)$ be the set of sequences of type μ in which for each i, the number of good i's is <u>at least</u> $\mu_i^{\#}$.

We write 0 for the partition of 0, so that $s(0,\mu)$ consists of all sequences of type μ. Since the number of good (i+1)'s in any sequence is at most the number of good i's there has been no loss in assuming that $\mu_{i+1}^{\#} \le \mu_i^{\#}$.

<u>15.7</u> If $\lambda_1^{\#} = \mu_1$ and $\lambda_i^{\#} = \mu_i^{\#}$ for $i > 1$, then $s(\lambda^{\#},\mu) = s(\mu^{\#},\mu)$, since every 1 in a sequence is good.

Thus we can absorb the first part of μ into $\mu^{\#}$.

15.8 EXAMPLE $s(0,(3,2)) = s((3),(3,2))$. The sequences in the second and third columns below give $s((3,1),(3,2))$ and the sequences in the last column give $s((3,2),(3,2))$.

$s((3),(3,2))$	$\supset$	$s((3,1),(3,2))$	$\supset$	$s((3,2),(3,2))$
2 2 1 1 1		2 1 2 1 1		1 2 1 2 1
		2 1 1 2 1		1 2 1 1 2
		2 1 1 1 2		1 1 2 2 1
		1 2 2 1 1		1 1 2 1 2
				1 1 1 2 2

Compare Example 5.2, where $M^{(3,2)}$ has a series of submodules with the factors of dimensions 1,4 and 5. This is no coincidence!

Given a pair $\mu^{\#}$, μ of partitions, we record them in a picture similar to a diagram. We shall draw a line between each row and enclose $\mu^{\#}$ by vertical lines. The picture for $\mu^{\#}$, μ will always be identified with the picture obtained by enclosing all the nodes in the first row (cf. 15.7).

15.9 EXAMPLE Referring to Example 15.8, we have

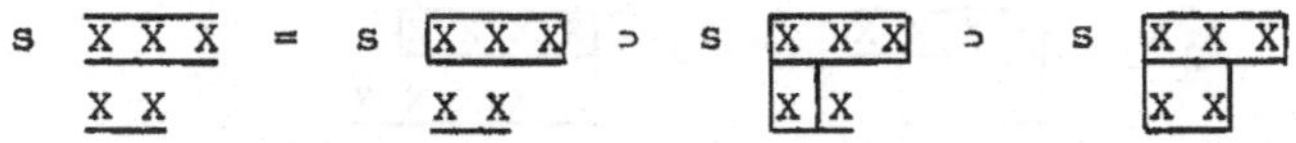

```
s  X X X   =   s  |X X X|   ⊃   s  |X X X|   ⊃   s  |X X X|
   X X            X X             |X|X             |X X|
```

This nesting suggests that we should have some notation which adds a node from μ to $\mu^{\#}$. We need only consider absorbing a node which is not in the first row.

15.10 DEFINITION Suppose $\mu^{\#} \neq \mu$. Let c be an integer greater than 1 such that $\mu^{\#}_c < \mu_c$ and $\mu^{\#}_{c-1} = \mu_{c-1}$.

(i) If $\mu^{\#}_{c-1} > \mu^{\#}_c$, then $\mu^{\#} A_c$, μ is the pair of partitions obtained from $\mu^{\#}, \mu$ by changing $\mu^{\#}_c$ to $\mu^{\#}_c + 1$. If $\mu^{\#}_{c-1} = \mu^{\#}_c$, then $\mu^{\#} A_c$, μ is the pair 0,0.

(ii) $\mu^{\#}, \mu R_c$ is the pair of partitions obtained from $\mu^{\#}, \mu$ by changing μ_c to $\mu^{\#}_c$ and μ_{c-1} to $\mu_{c-1} + \mu_c - \mu^{\#}_c$.

The operator R_c merely moves some nodes lying outside $\mu^{\#}$ to the end of the row above (R stands for "raise" and A stands for "add"). Both $\mu^{\#}$ and μ are involved in the definitions of A_c and R_c, and note that we stipulate that $\mu^{\#}_{c-1}$ equals μ_{c-1}.

15.11 EXAMPLE

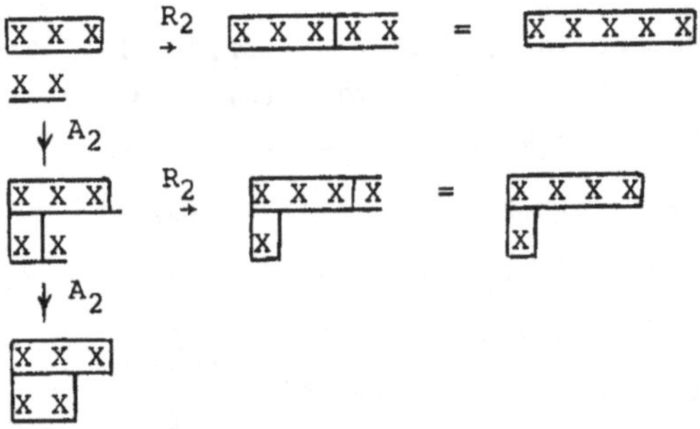

```
[X X X]   R2   [X X X|X X]   =   [X X X X X]
 X X       →
   ↓ A2
[X X X]   R2   [X X X|X]     =   [X X X X]
[X|X       →   [X]               [X]
   ↓ A2
[X X X]
[X X]
```

Other examples are given in 15.13, 17.15 and 17.16.

Since R_c raises some nodes, and we always enclose all the nodes in the first row, any sequence of operations A_c, R_c on a pair of partitions leads eventually to a pair of partitions of the form λ, λ (when, perforce, λ is a proper partition.) It is also clear that

<u>15.12</u> Given any pair of partitions, $\mu^{\#}, \mu$, there is a partition ν and a sequence of operations A_c, R_c leading from $0, \nu$ to $\mu^{\#}, \mu$.

15.13 EXAMPLE To obtain $((4,3,1),(4,5,2^2))$, apply $A_2^3\, A_3\, R_4\, R_3\, R_5\, R_6\, R_4\, R_5$ to $(0,(4,3,1,2,1,2))$:

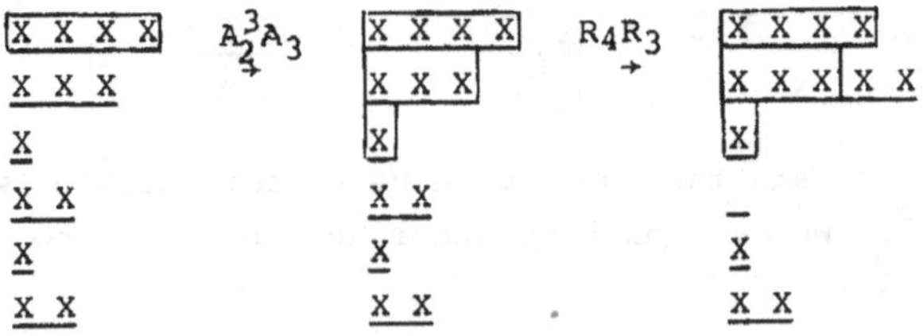

```
[X X X X]   A2^3 A3   [X X X X]   R4R3   [X X X X]
 X X X         →      [X X X]      →     [X X X|X X
 X                    [X]                [X]
 X X                   X X                _
 X                     X                  X
 X X                   X X                X X
```

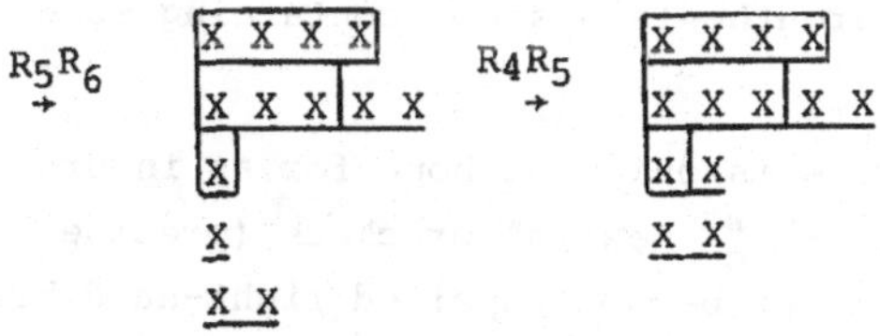

$\xrightarrow{R_5R_6}$

X X X X
X X X X X
X
X
X X

$\xrightarrow{R_4R_5}$

X X X X
X X X X X
X X
X X

The critical theorem for sequences is

15.14 THEOREM <u>The following gives a 1-1 correspondence between $s(\mu^{\#},\mu) \setminus s(\mu^{\#}A_c,\mu)$ and $s(\mu^{\#},\mu R_c)$:</u>

<u>Given a sequence in the first set, change all the bad c's to (c-1)'s.</u>

<u>Proof</u>: Recall that our definition of the operators A_c and R_c required $\mu_{c-1} = \mu^{\#}_{c-1}$. Therefore, a sequence s_1 in $s(\mu^{\#},\mu) \setminus s(\mu^{\#}A_c,\mu)$ contains

$\mu_{c-1} = \mu^{\#}_{c-1}$ (c-1)'s, all good.

$\mu^{\#}_c$ good c's and $\mu_c - \mu^{\#}_c$ bad c's.

The bad c's are changed to (c-1)'s to give a sequences s_2. We claim that

<u>15.15</u> For all j, the number of good (c-1)'s before the jth term of $s_2 \geq$ the number of good (c-1)'s before the jth term in s_1.

This is certainly true for j = 1, so assume true for j = i. Then 15.15 is obviously true for j = i + 1, except when the ith term is a (c-1) which is good in s_1 but bad in s_2. But in this case, the inequality in 15.15 (with j replaced by i) is strict, because the number of (c-2)'s before the ith term is the same in both s_1 and s_2. Therefore, 15.15 is true for j = i + 1 in this case also.

15.15 shows that s_2 has at least $\mu^{\#}_{c-1}$ good (c-1)'s, and that all the c's in s_2 are good. Hence, for $i \neq c-1$ or c, i is good in s_2 if and only if i is good in s_1, and so s_2 belongs to $s(\mu^{\#},\mu R_c)$.

It is more difficult to prove the given map 1-1 and onto.

Given any sequence replace all the (c-1)'s by left-hand brackets, (, and all the c's by right-hand brackets,). For example, if c = 3

1 2 1 2 3 1 2 3 3 2 2 1 1 3 1 1 2 2 3

goes to

1 (1 ( ) 1 ( )) ((1 1) 1 1 (( ) .

Now, in any sequence belonging to $s(\mu^{\#},\mu R_c)$, all the c's are good. Therefore, every right-hand bracket is preceded by more left-hand brackets than right-hand brackets, and the sequence looks like

$$p_0(p_1(p_2(\ldots(p_r \quad \text{with} \quad r = \mu_{c-1} + \mu_c - 2\mu^{\#}_c,$$

where each p_j is a closed parenthesis system, containing some terms i with $i \neq c-1$ or c.

It is now clear that there is only one hope for an inverse map; namely, reverse the first $\mu_c - \mu_c^{\#}$ "extra" brackets (precisely the brackets which are reversed must become unpaired right-hand brackets, to give us an inverse image.)

Let s belong to $s(\mu^{\#}, \mu R_c)$. We say that a c-1 is <u>black</u> in s if it corresponds to an extra bracket; otherwise it is <u>white</u>.

Let s^* be the sequence obtained from s by changing the first $\mu_c - \mu_c^{\#}$ black (c-1)'s to c's. We must prove

<u>15.16</u> Every c-1 in s^* is good.

The Theorem will then follow, since every c appearing in both s and s^* will be good, and s^* will be the unique element of $s(\mu^{\#}, \mu)$ $s(\mu^{\#} A_c, \mu)$ mapping to s.

We tackle the proof of 15.16 in two steps. First

<u>15.17</u> For every term x in s, the number of white (c-1)'s before x $\leq$ the number of good (c-1)'s before x.

Initially, let x be a black c-1. The number of white (c-1)'s before x = the number of c's before x (by the definition of "black") $\leq$ the number of good (c-1)'s before x, since every c in s is good. This proves 15.17 in the case where x is a black c-1.

The same proof shows that the number of white (c-1)'s in s $\leq$ the number of good (c-1)'s in s. Thus, we may start at the end of s and work back, noting that 15.17 is trivially true for the (j-1)th term of s if it is true for the jth term, except when the (j-1)th term is a black c-1, which is the case we have already done.

Next we have

<u>15.18</u> Either c = 2, or for every c-1 in s^*, the number of previous good (c-2)'s > the number of previous (c-1)'s in s^*.

For the proof of 15.18, assume c > 2. Now, s contains at most $\mu_c - \mu_c^{\#}$ bad (c-1)'s since s belongs to $s(\mu^{\#}, \mu R_c)$, so for any c-1 in s, the number of previous good (c-2)'s > the number of previous (c-1)'s in s - $(\mu_c - \mu_c^{\#})$. Therefore, 15.18 holds for a c-1 after the $(\mu_c - \mu_c^{\#})$th black c-1 in s.

If the term x in s^* is a c-1 appearing before the $(\mu_c - \mu_c^{\#})$th black c-1 in s, then x was white in s. Also, the number of (c-1)'s before x in s^* = the number of white (c-1)'s before x in s $\leq$ the number of good (c-1)'s before x in s by 15.17 (the inequality being strict if x is a bad c-1 in s, by applying 15.17 to the next term) ≤ the number of good (c-2)'s before x (the inequality being strict if x is a good c-1 in s), and 15.18 is proved in this case too.

From 15.18, 15.16 follows at once, and this completes the proof of Theorem 15.14.

15.19 EXAMPLE Referring to Example 15.8, the 1-1 correspondence between s((3),(3,2)) \ s((3,1),(3,2)) and s((5),(5)) is obtained by:

```
2 2 1 1 1  →  1 1 1 1 1
× × ✓ ✓ ✓
```

The 1-1 correspondence between s((3,1),(3,2)) \ s((3,2),(3,2)) and s((4,1),(4,1)) is given by

```
2 1 2 1 1        1 1 2 1 1
× ✓ ✓ ✓ ✓
2 1 1 2 1        1 1 1 2 1
× ✓ ✓ ✓ ✓
             →
2 1 1 1 2        1 1 1 1 2
× ✓ ✓ ✓ ✓
1 2 2 1 1        1 2 1 1 1
✓ ✓ × ✓ ✓
```

16 THE LITTLEWOOD-RICHARDSON RULE

The Littlewood-Richardson Rule is an algorithm for calculating $[\lambda][\mu]$ where λ is a proper partition of n-r and μ is a proper partition of r. Remember that $[\lambda][\mu]$ is a linear combination of diagrams with n nodes, and the interpretation is that when a_ν is the coefficient of $[\nu]$, $S^\lambda_{\mathbb{Q}} \otimes S^\mu_{\mathbb{Q}} \uparrow \mathfrak{S}_n$ has $S^\nu_{\mathbb{Q}}$ as a composition factor with multiplicity a_ν. It is a well-known result that every ordinary irreducible representation of G × H, for groups G and H is equivalent to $S_1 \times S_2$, for some irreducible G-module S_1 and some irreducible H-module S_2, so the Littlewood-Richardson Rule enables us to calculate the composition factors of any ordinary representation of a Young subgroup, induced up to $\mathfrak{S}_n$.

For the moment, forget any intended interpretation, and consider the additive group generated by $\{[\lambda] \mid \lambda$ is a proper partition of some integer$\}$. Given any pair of partitions $\mu^\#, \mu$ as in definition 15.5, we define a group endomorphism $[\mu^\#, \mu]^\bullet$ of this additive group as follows:

<u>16.1</u> DEFINITION $[\lambda]^{[\mu^\#,\mu]^\bullet} = \Sigma\, a_\nu[\nu]$ where $a_\nu = 0$ unless $\lambda_i \le \nu_i$ for every i, and if $\lambda_i \le \nu_i$ for every i, then a_ν is the number of ways of replacing the nodes of $[\nu]\backslash[\lambda]$ by integers such that

(i) The numbers are non-decreasing along rows

and (ii) The numbers are strictly increasing down columns

and (iii) When reading from <u>right to left</u> in successive rows, we have a sequence in $s(\mu^\#, \mu)$.

If $\mu^\# = \mu$, when μ is <u>a fortiori</u> a proper partition, we write $[\mu]^\bullet$ for $[\mu,\mu]^\bullet$.

The operators are illustrated by the next Lemma and by Examples 16.6 and 16.7.

16.2 LEMMA <u>If $\mu = (\mu_1, \mu_2, \ldots, \mu_k)$, then $[0]^{[0,\mu]^\bullet} = [\mu_1][\mu_2]\ldots[\mu_k]$. If μ is a proper partition, then $[0]^{[\mu]^\bullet} = [\mu]$.</u>

<u>Proof</u>: When $\mu^\# = 0$, condition (iii) of definition 16.1 merely says that we have a sequence of type μ. But $[\mu_1][\mu_2]\ldots[\mu_k]$, by definition, describes the composition factors of $M^\mu_{\mathbb{Q}}$, and the first result follows from Young's Rule.

Let $[\nu]$ be a diagram appearing in $[0]^{[\mu]^\bullet}$. Then the nodes in $[\nu]$ can be replaced by μ_1 1's, μ_2 2's, and so on, in such a way that conditions (i) to (iii) of 16.1 hold. Suppose that some i appears in the jth row with j < i, and let i be the least number for which this happens. There are no (i-1)'s higher than this i, by the minimality of of i; nor can there be any (i-1)'s to the right of it in the same row, by condition (i). Thus, this i is preceded by no (i-1)'s when reading from right to left in successive rows, and the i is bad, contradicting condition (iii). But no i can appear in the jth row with $j > i$, by condition (ii). This proves that every i is in the ith row, and $[\nu] = [\mu]$.

16.3 LEMMA $\underline{[\mu^{*},\mu]^{\cdot} = [\mu^{*}A_c,\mu]^{\cdot} + [\mu^{*},\mu R_c]^{\cdot}}$

<u>Proof</u>: Assume that μ is a partition of r, and that λ and ν are proper partitions of n-r and n, respectively, with $\lambda_i \le \nu_i$ for each i. Replace each node in $[\nu]\backslash[\lambda]$ by μ_1 1's, μ_2 2's and so on, such that we have a sequence in $s(\mu^{*},\mu)\setminus s(\mu^{*}A_c,\mu)$ when reading from right to left in successive rows. We must prove that changing all the bad c's to (c-1)'s gives a configuration of integers satisfying 16.1 (i) and (ii) if and only if we start with a configuration of integers satisfying 16.1 (i) and (ii), since the Lemma will then follow from Theorem 15.14.

Suppose we have not yet changed the bad c's to (c-1)'s and conditions 16.1 (i) and (ii) hold for our configuration of integers. There are two problems which might occur. A bad c might be to the right of a good c in the same row. This cannot happen, because a c immediately after a bad c must itself be bad. More complicated is the possibility that there is a bad c in the (i,j)th place and a c-1 in the (i-1,j)th place. To deal with this, let m be maximal such that there are c's in the (i,j)th,(i,j+1)th,...,(i,j+m-1)th places. Then by conditions 16.1 (i) and (ii), there are (c-1)'s in the (i-1,j)th,(i-1,j+1)th,...,(i-1, j+m-1)th places. Since all the (c-1)'s are good in a sequence belonging to $s(\mu^{*},\mu)\setminus s(\mu^{*}A_c,\mu)$, our c in the (i,j)th place must be good, after all, by 15.4. This shows that all the bad c's can be changed to (c-1)'s without affecting conditions 16.1(i) and (ii).

Conversely, suppose that after changing the bad c's to (c-1)'s we have a configuration satisfying conditions 16.1 (i) and (ii). We discuss the configuration of integers we started with. This must satisfy conditions 16.1 (i) and (ii) unless a bad c occurs immediately to the left of a (good) c-1 in the same row, or a bad c lies immediately above a good c in the same column. The first problem cannot occur by 15.4. Therefore, we have only to worry about the possibility that a bad c is in the (i-1,j)th place and a good c is in the (i,j)th place. Reading from right to left in successive rows, we see that the number of (good) (c-1)'s in the (i-1)th row to the left of our bad c in the (i-1,j)th place is at least the number of good c's in the ith row. But every c-1 in the (i-1)th row to the left of the (i-1,j)th place must have a good c immediately below it in the ith row (since there is a good c in the (i,j)th place, and we end up with a configuration satisfying conditions 16.1 (i) and (ii)). This contradicts the fact that there is a good c in the (i,j)th place, and completes the proof of the Lemma.

16.4 THE LITTLEWOOD-RICHARDSON RULE

$$[\lambda]^{[\mu]^{\bullet}} = [\lambda][\mu]$$

Proof: (James [10]) If ν is a proper partition of n, then applying operators A_c and R_c repeatedly to O, ν we reach a collection of pairs of partitions ω, ω. By Lemma 16.3, we may write

$$[O,\nu]^{\bullet} = \sum_{\omega} a_{\omega}[\omega]^{\bullet}$$

where each a_{ω} in an integer, $a_{\nu} = 1$ and $a_{\omega} = 0$ unless $[\omega] \trianglerighteq [\nu]$.

Hence there are integers b_{α} and c_{β} such that

$$[\lambda]^{\bullet} = \sum_{\alpha} b_{\alpha}[O,\alpha]^{\bullet} \text{ and } [\mu]^{\bullet} = \sum_{\beta} c_{\beta}[O,\beta]^{\bullet}.$$

By Lemma 16.2

$$\begin{aligned}[\lambda]^{[\mu]^{\bullet}} &= [O]^{[\lambda]^{\bullet}[\mu]^{\bullet}} \\ &= [O]^{\Sigma b_{\alpha}[O,\alpha]^{\bullet}\Sigma c_{\beta}[O,\beta]^{\bullet}} \\ &= \sum_{\alpha} b_{\alpha}\ [\alpha_1]\ldots[\alpha_j] \sum_{\beta} c_{\beta}\ [\beta_1]\ldots[\beta_k] \\ &= [O]^{\Sigma b_{\alpha}[O,\alpha]^{\bullet}}[O]^{\Sigma c_{\beta}[O,\beta]^{\bullet}} \\ &= [O]^{[\lambda]^{\bullet}}[O]^{[\mu]^{\bullet}} \\ &= [\lambda][\mu]\ .\end{aligned}$$

16.5 COROLLARY $[\nu]^{\bullet}[\mu]^{\bullet} = [\mu]^{\bullet}[\nu]^{\bullet} = ([\mu][\nu])^{\bullet}$

Proof: For all $[\lambda]$, $[\lambda]^{[\nu]^{\bullet}[\mu]^{\bullet}} = [\lambda][\nu][\mu] = [\lambda][\mu][\nu]$

$$= [\lambda]^{[\mu]^{\bullet}[\nu]^{\bullet}} = [\lambda]^{([\mu][\nu])^{\bullet}}$$

The Corollary is extremely hard to prove directly. More generally, it follows from the Littlewood-Richardson Rule that for every equation like $[3][2] = [5] + [4,1] + [3,2]$ there is a corresponding operator equation $[3]^{\bullet}[2]^{\bullet} = [5]^{\bullet} + [4,1]^{\bullet} + [3,2]^{\bullet}$.

Of course, the Branching Theorem (part (i)) is a special case of the Littlewood-Richardson Rule.

When applying the Littlewood-Richardson Rule, it is best to draw the diagram $[\lambda]$, then add μ_1 1's, then μ_2 2's and so on, making sure that at each stage $[\lambda]$, together with the numbers which have been added, form a proper diagram shape and no two identical numbers appear in the same column. Then reject the result unless reading from right to left in successive rows each i is preceded by more (i-1)'s than i's. (This condition is necessary and sufficient for every term to be good.)

16.6 EXAMPLE $[3,2][2] = [3,2]^{[2]^{\bullet}}$

$= [5,2] + [4,3] + [4,2,1] + [3^2,1] + [3,2^2]$, by looking at the following configurations: (cf. Example 14.5).

```
X X X 1 1    X X X 1    X X X 1    X X X    X X X
X X          X X 1      X X        X X 1    X X
                        1          1        1 1
```

16.7 EXAMPLE $[3,2][2][2] = [3,2]^{[2]^{\bullet}[2]^{\bullet}} =$

```
X X X 1 1    X X X 1 1    X X X 1 1    X X X 1    X X X 1
X X 2 2      X X 2        X X          X X 1 2    X X 1
             2            2 2          2          2 2

X X X 1      X X X 1      X X X        X X X
X X 2        X X          X X 1        X X
1            1 2          1 2          1 1
2            2            2            2 2
```

```
X X X 1 1 2    X X X 1 1 2    X X X 1 2    X X X 1 2
X X 2          X X            X X 1 2      X X 1
               2                           2

X X X 1 2    X X X 1 2    X X X 1 2    X X X 1    X X X 1
X X 2        X X          X X          X X 2 2    X X 2
1            1 2          1            1          1 2
                          2

X X X 2      X X X 2      X X X        X X X 2    X X X
X X 1        X X 1        X X 1        X X        X X 2
1 2          1            1 2 2        1 1        1 1
             2                         2          2
```

```
X X X 1 1 2 2    X X X 1 2 2    X X X 1 2 2    X X X 2 2
X X              X X 1          X X            X X 1
                                1              1

X X X 2 2    X X X 2
X X          X X 2
1 1          1 1
```

We have arranged the diagrams so that, reading from right to left in successive rows, the diagrams in the first batch (before the first line) give sequences in s((2,2),(2,2)), so

$$[3,2][2,2] = [3,2]^{[2,2]^{\bullet}} = [5,4] + [5,3,1] + [5,2^2] + [4^2,1] + [4,3,2] + [4,3,1^2] + [4,2^2,1] + [3^2,2,1] + [3,2^3]$$

The diagrams before the second line give $[3,2]^{[(2,1),(2,2)]^{\bullet}}$. The reader may care to check that changing a bad 2 to a 1 in the second batch gives the same answer as $[3,2]^{[3,1]^{\bullet}}$, in agreement with Lemma 16.3.

$$[3,2][3,1] = [3,2]^{[3,1]^{\bullet}} = [6,3] + [6,2,1] + [5,4] + 2[5,3,1] + [5,2^2] + [5,2,1^2] + [4^2,1] + 2[4,3,2] + [4,3,1^2] + [3^3] + [4,2^2,1]\ [3^2,2,1].$$

The last batch contains all the configurations where both 2's are bad, and by changing the 2's to 1's, Lemma 16.3 gives

$$[3,2][4] = [3,2]^{[4]^{\bullet}} = [7,2] + [6,3] + [6,2,1] + [5,3,1] + [5,2^2] + [4,3,2] ,$$

which is simple to verify directly.

17. A SPECHT SERIES FOR M^μ

A better form of Young's Rule can be derived over an arbitrary field. What happens in this case is that M^μ has a series with each factor isomorphic to a Specht module; such a series will be called a <u>Specht series</u>. Since M^μ is not completely reducible over some fields, we must take into account the order of the factors in a Specht series. The next example shows that the order of the factors does matter:

17.1 EXAMPLE Let char F divide $n > 2$, and consider $M^{(n-1,1)}$. Example 5.1 shows that $M^{(n-1,1)}$ is uniserial, with factors $D^{(n)}, D^{(n-1,1)}$ $D^{(n)}$ and that $S^{(n-1,1)}$ is uniserial with factors $D^{(n-1,1)}, D^{(n)}$, reading from the top. Thus $M^{(n-1,1)}$ has no Specht series with factors $S^{(n-1,1)}$, $S^{(n)}$ reading from the top. The Specht series in Example 5.1 has factors in the order $S^{(n)}, S^{(n-1,1)}$.

In this important section, we shall use only Theorem 15.14 on sequences, and deduce both Young's Rule and the standard basis of the Specht module. At the same time, we characterize the Specht module S^λ as the intersection of certain $F\mathfrak{S}_n$-homomorphisms defined on M^λ, in the case where λ is a proper partition. Throughout this section F is an arbitrary field.

Let $\mu^\#, \mu$ be a pair of partitions for n, as defined in 15.5. Recall that $\mu^\#$ must be a proper partition, while we do not require μ to be proper. We want to define a submodule $S^{\mu^\#,\mu}$ of M^μ, and to do this we construct an object $e_t^{\mu^\#,\mu}$ which is "between" a tabloid and a polytabloid.

17.2 DEFINITION Suppose that t is a μ-tableau. Let

$$e_t^{\mu^\#,\mu} = \Sigma\ \{\operatorname{sgn} \pi\{t\}\pi \mid \pi \in C_t \text{ and } \pi \text{ fixes the numbers outside } [\mu^\#]\}$$

17.3 EXAMPLE If $t = \begin{array}{llll} 1 & 3 & 5 & \\ 2 & 7 & 4 & 9 \\ 8 & 6 & & \end{array}$ and $\mu^\# = (3,2,0)$, $\mu = (3,4,2)$

(part of t is boxed-in only to show which numbers will be moved), then

$$e_t^{\mu^\#,\mu} = \begin{array}{l} \overline{\underline{1\ 3\ 5}} \\ \underline{2\ 7\ 4\ 9} \\ \underline{8\ 6} \end{array} - \begin{array}{l} \overline{\underline{2\ 3\ 5}} \\ \underline{1\ 7\ 4\ 9} \\ \underline{8\ 6} \end{array} - \begin{array}{l} \overline{\underline{1\ 7\ 5}} \\ \underline{2\ 3\ 4\ 9} \\ \underline{8\ 6} \end{array} + \begin{array}{l} \overline{\underline{2\ 7\ 5}} \\ \underline{1\ 3\ 4\ 9} \\ \underline{8\ 6} \end{array}$$

17.4 DEFINITION $S^{\mu^\#,\mu}$ is the subspace of M^μ spanned by $e_t^{\mu^\#,\mu}$'s as t varies.

Of course, $S^{\mu^\#,\mu}$ is an $F\mathfrak{S}_n$-submodule of M^μ, since $e_t^{\mu^\#,\mu}\pi = e_{t\pi}^{\mu^\#,\mu}$

If $\mu^{\#} = 0$, then $S^{\mu^{\#},\mu} = M^{\mu}$ and if $\mu^{\#} = \mu$, then $S^{\mu^{\#},\mu} = S^{\mu}$.

17.5 If $\lambda_1^{\#} = \mu_1$ and $\lambda_i^{\#} = \mu_i^{\#}$ for $i > 1$, then $S^{\lambda^{\#},\mu} = S^{\mu^{\#},\mu}$, so we can absorb the first part of μ into $\mu^{\#}$ (cf. 15.7).

Sequences now come into play by way of

17.6 CONSTRUCTION Given a sequence of type μ, construct a corresponding μ-tableau t as follows. Work along the sequence. If the jth term is a good i, put j as far left in the ith row of t as possible. If the jth term is a bad i, put j as far right in the ith row as possible.

17.7 EXAMPLE
$$\begin{matrix} 3 & 1 & 1 & 2 & 3 & 3 & 2 & 3 & 2 & 1 & 2 & 1 \\ \times & \checkmark & \checkmark & \checkmark & \checkmark & \times & \checkmark & \checkmark & \times & \checkmark & \checkmark & \checkmark \end{matrix} \in s((4,3,2),(4,4,4))$$

and corresponds to
$$\begin{matrix} 2 & 3 & 10 & 12 \\ 4 & 7 & 11 & 9 \\ 5 & 8 & 6 & 1 \end{matrix}$$

Different sequences in $s(0,\mu)$ correspond to tableaux which belong to different tabloids, so

17.8 The construction gives a 1-1 correspondence between $s(0,\mu)$ and the set of μ tabloids.

Remark We have already encountered the concept of viewing a basis of M^{μ} as a set of sequences, for in section 13, the tableau T of type μ corresponds to the sequence $(1)T, (2)T, \ldots, (n)T$.

The construction ensures that a sequence in $s(\mu^{\#},\mu)$ corresponds to a tableau which is standard inside $[\mu^{\#}]$ (i.e. the numbers which lie inside $[\mu^{\#}]$ increase along rows and down columns- cf. Example 17.7). But, if t is standard inside $[\mu^{\#}]$, then $\{t\}$ is the last tabloid involved in $e_t^{\mu^{\#},\mu}$ (cf. Example 17.3), and so Lemma 8.2 gives

17.9 $\{ e_t^{\mu^{\#},\mu} \mid t$ corresponds to a sequence in $s(\mu^{\#},\mu)$ by 17.6$\}$ is a linearly independent subset of $S^{\mu^{\#},\mu}$.

We shall see soon that we actually have a basis of $S^{\mu^{\#},\mu}$ here. Our main objective, though, is to prove that $S^{\mu^{\#},\mu}/S^{\mu^{\#}A_c,\mu} \cong S^{\mu^{\#},\mu R_c}$, where the operators A_c and R_c are defined in 15.10. First, note that $S^{\mu^{\#}A_c,\mu} \subseteq S^{\mu^{\#},\mu}$. This is trivially true if $\mu^{\#}A_c,\mu = 0,0$ (i.e. if $\mu_{c-1}^{\#} = \mu_c^{\#}$), since we adopt the convention that $S^{0,0}$ is the zero module. Otherwise, given t, we may take coset representatives $\sigma_1,\ldots,\sigma_k$ for the subgroup of C_t fixing the numbers outside $[\mu^{\#}]$ in the subgroup of

C_t fixing the numbers outside $[\mu^* A_c]$, whereupon $e_t^{\mu^* A_c,\mu} = e_t^{\mu^*,\mu} \sum_{i=1}^{k} (\operatorname{sgn} \sigma_i)\sigma_i$.

Now we want an $F\mathfrak{S}_n$-homomorphism mapping $S^{\mu^*,\mu}$ onto $S^{\mu^*,\mu R_c}$.

17.10 DEFINITION Let $\mu = (\mu_1,\mu_2,\ldots)$ and $\nu = (\mu_1,\mu_2,\ldots,\mu_{i-1},\mu_i + \mu_{i+1} - v, v, \mu_{i+2},\ldots)$. Then $\psi_{i,v}$ belonging to $\operatorname{Hom}_{F\mathfrak{S}_n}(M^\mu,M^\nu)$ is defined by $\{t\}\psi_{i,v} = \Sigma\ \{\{t_1\} \mid \{t_1\}$ agrees with $\{t\}$ on all except the ith and $(i+1)$th rows, and the $(i+1)$th row of $\{t_1\}$ is a subset of size v in the $(i+1)$th row of $\{t\}\}$.

Remark It is slightly simpler to visualize the action of $\psi_{i,v}$ on the basis of M^μ viewed as sequences. $\psi_{i,v}$ sends a sequence to the sum of all sequences obtained by changing all but v $(i+1)$'s to i's. Whichever way you look at it, $\psi_{i,v}$ is obviously an $F\mathfrak{S}_n$-homomorphism. Every tabloid involved in $\{t\}\psi_{i,v}$ has coefficient 1, so $\psi_{i,v}$ is "independent of the ground field."

17.11 EXAMPLES

(i) When $\mu = (3,2)$, $\psi_{1,0}$ and $\psi_{1,1}$ are the homomorphisms ψ_0 and ψ_1 appearing in Example 5.2.

(ii) If $\mu = (4,3^2,2)$, then

$$\psi_{2,1}: \quad \begin{array}{l}\overline{1\ 2\ 5\ 10}\\ \overline{3\ 4\ 9}\\ \overline{6\ 7\ 8}\\ \overline{11\ 12}\end{array} \to \begin{array}{l}\overline{1\ 2\ 5\ 10}\\ \overline{3\ 4\ 9\ 7\ 8}\\ \overline{6}\\ \overline{11\ 12}\end{array} + \begin{array}{l}\overline{1\ 2\ 5\ 10}\\ \overline{3\ 4\ 9\ 6\ 8}\\ \overline{7}\\ \overline{11\ 12}\end{array} + \begin{array}{l}\overline{1\ 2\ 5\ 10}\\ \overline{3\ 4\ 9\ 6\ 7}\\ \overline{8}\\ \overline{11\ 12}\end{array}$$

(iii) If $n \geq 6$ and $\mu = (n-3,3)$ and $v = \overline{1\ 2\ 3} + \overline{1\ 2\ 4} + \overline{1\ 3\ 4} + \overline{2\ 3\ 4}$ (replacing each tabloid by its second row only), we have

$$v\,\psi_{1,0} = 4 \in F$$

$$v\,\psi_{1,1} = \bar{1} + \bar{2} + \bar{3} + \bar{1} + \bar{2} + \bar{4} + \bar{1} + \bar{3} + \bar{4} + \bar{2} + \bar{3} + \bar{4}$$

$$= 3(\bar{1} + \bar{2} + \bar{3})$$

$$v\,\psi_{1,2} = 2(\overline{1\ 2} + \overline{1\ 3} + \overline{1\ 4} + \overline{2\ 3} + \overline{2\ 4} + \overline{3\ 4}).$$

Therefore, $v \in \operatorname{Ker}\psi_{1,0} \cap \operatorname{Ker}\psi_{1,2}$ if and only if char $F = 2$

and $v \in \operatorname{Ker}\psi_{1,1}$ if and only if char $F = 3$.

(iv) Taking n = 6 in example (iii),

$$(\overline{4\ 5\ 6} - \overline{1\ 5\ 6})\psi_{1,1} = \bar{4} + \bar{5} + \bar{6} - \bar{1} - \bar{5} - \bar{6} = \bar{4} - \bar{1}$$

$$(\overline{4\ 5\ 6} - \overline{1\ 5\ 6} - \overline{4\ 2\ 6} + \overline{1\ 2\ 6})\psi_{1,1} = 0.$$

That is, for t = $\begin{matrix} 1 & 2 & 3 \\ 4 & 5 & 6 \end{matrix}$, $\mu^{\#} = (3,1)$ and $\mu = (3,3)$, we have

$$e_t^{\mu^{\#},\mu}\ \psi_{1,1} = e_{tR_2}^{\mu^{\#},\mu R_2} \quad \text{where } tR_2 = \begin{matrix} 1 & 2 & 3 & 5 & 6 \\ 4 & & & & \end{matrix}$$

and $$e_t^{\mu^{\#} A_2,\mu}\ \psi_{1,1} = 0.$$

Compare the last Example with

17.12 LEMMA $\underline{S^{\mu^{\#},\mu}\ \psi_{c-1,\mu_c^{\#}}} = \underline{S^{\mu^{\#},\mu R_c}}$

and $\underline{S^{\mu^{\#} A_c,\mu}\ \psi_{c-1,\mu_c^{\#}} = 0.}$

<u>Proof:</u> Let t be a μ-tableau, and let

$\kappa_{t^{\#}} = \Sigma\ \{(\text{sgn }\pi)\pi \mid \pi$ fixes the numbers in t outside $[\mu^{\#}]\}$.

Choose a set B of $\mu_c^{\#}$ numbers from the cth row of t, and move the rest of the numbers in the cth row of t into the (c-1)th row.

If B consists of the first $\mu_c^{\#}$ numbers in the cth row of t, then we get a tableau, tR_c say, and

$$\{tR_c\}\kappa_{t^{\#}} = e_{tR_c}^{\mu^{\#},\mu R_c}.$$

For any other set of $\mu_c^{\#}$ numbers from the cth row of t, we still get a μR_c -tabloid, $\{t_1\}$ say, but now one of the numbers, say x, which has been moved up lies inside $[\mu^{\#}]$. Let y be the number above x in t. Then $(1-(x\ y))$ is a factor of κ_t , and so

$$\{t_1\}\kappa_{t^{\#}} = 0.$$

Now, $e_t^{\mu^{\#},\mu}\ \psi_{c-1,\mu_c^{\#}} = \{t\}\kappa_{t^{\#}}\psi_{c-1,\mu_c^{\#}} = \{t\}\ \psi_{c-1,\mu_c^{\#}}\ \kappa_{t^{\#}}$ and

$\{t\}\ \psi_{c-1,\mu_c^{\#}}$ is the sum of all the tabloids obtained by moving all except $\mu_c^{\#}$ numbers from the cth row of $\{t\}$ into the (c-1)th row. Therefore,

$$e_t^{\mu^{\#},\mu}\ \psi_{c-1,\mu_c^{\#}} = e_{tR_c}^{\mu^{\#},\mu R_c}.$$

Since $\mu^{\#} A_c,\mu$ has one more node enclosed in the cth row (or $S^{\mu^{\#}A_c,\mu} = S^{0,0} = 0$ if $\mu_{c-1}^{\#} = \mu_c^{\#}$), the proof we used to deduce that $\{t_1\}\kappa_{t^{\#}} = 0$ shows that $e_t^{\mu^{\#}A_c,\mu}\psi_{c-1,\mu_c^{\#}} = 0$.

17.13 THEOREM (James [10])

(i) $S^{\mu^\#,\mu}\psi_{c-1,\mu_c^\#} = S^{\mu^\#,\mu R_c}$ and

$S^{\mu^\#,\mu} \cap \ker \psi_{c-1,\mu_c^\#} = S^{\mu^\# A_c,\mu}$

(ii) $S^{\mu^\#,\mu}/S^{\mu^\# A_c,\mu} \cong S^{\mu^\#,\mu R_c}$

(iii) $\dim S^{\mu^\#,\mu} = |s(\mu^\#,\mu)|$; indeed, $\{e_t^{\mu^\#,\mu} \mid t$ corresponds to a sequence in $s(\mu^\#,\mu)$ by 17.6$\}$ is a basis of $S^{\mu^\#,\mu}$.

(iv) $S^{\mu^\#,\mu}$ has a Specht series. The factors in this series are given by $[O]^{[\mu^\#,\mu]^\bullet}$.

Proof: Let O,ν be a pair of partitions from which we can reach the pair $\mu^\#,\mu$ by a sequence of A_c and R_c operators (cf. 15.12)

$\dim S^{O,\nu} = \dim M^\nu = |s(O,\nu)|$ by 17.8. We may therefore assume that $\dim S^{\mu^\#,\mu} = |s(\mu^\#,\mu)|$ and prove that $\dim S^{\mu^\# A_c,\mu} = |s(\mu^\# A_c,\mu)|$ and $\dim S^{\mu^\#,\mu R_c} = |s(\mu^\#,\mu R_c)|$.

Now,
$$\begin{aligned} |s(\mu^\#,\mu)| &= \dim S^{\mu^\#,\mu} \\ &\ge \dim S^{\mu^\# A_c,\mu} + \dim S^{\mu^\#,\mu R_c} \text{ by Lemma 17.12} \\ &\ge |s(\mu^\# A_c,\mu)| + |s(\mu^\#,\mu R_c)| \text{ by 17.9} \\ &= |s(\mu^\#,\mu)| \text{ by Theorem 15.14}. \end{aligned}$$

Everything falls out! We must have equality everywhere, so results (i), (ii) and (iii) follow.

When $\mu^\# = \mu$, $S^{\mu^\#,\mu} = S^\mu$, and so has a Specht series whose factors are given by $[O]^{[\mu]^\bullet}$ (see Lemma 16.2). Therefore, we may assume inductively that $S^{\mu^\# A_c,\mu}$ and $S^{\mu^\#,\mu R_c}$ have Specht series whose factors are given by $[O]^{[\mu^\# A_c,\mu]^\bullet}$ and $[O]^{[\mu^\#,\mu R_c]^\bullet}$. Since we have proved conclusion (i), and $[\mu^\#,\mu]^\bullet = [\mu^\# A_c,\mu]^\bullet + [\mu^\#,\mu R_c]^\bullet$ (see Lemma 16.3), $S^{\mu^\#,\mu}$ has a Specht series whose factors are given by $[O]^{[\mu^\#,\mu]^\bullet}$.

All we have used in the above proof are the purely combinatorial results 15.14 and 16.3 (In fact, it is much easier to show that $[O]^{[\mu^\#,\mu]^\bullet} = [O]^{[\mu^\# A_c,\mu]^\bullet + [\mu^\#,\mu R_c]^\bullet}$ than to prove Lemma 16.3 in its full form.) We have therefore given alternative proofs that the standard polytabloids form a basis for the Specht module (take $\mu^\# = \mu$ in part (iii)), and of Young's Rule (take $\mu^\# = O$ in part (iv)).

17.14 COROLLARY M^μ has a Specht series. More generally, $S^\lambda \otimes S^{(\mu_1)} \otimes \ldots \otimes S^{(\mu_k)} \uparrow \mathfrak{S}_n$ has a Specht series. The factors and their order of appearance are independent of the ground field, and can be calculated by applying the operators A_c and R_c repeatedly to $[0,\mu]$ and $[\lambda,(\lambda,\mu_1,\ldots,\mu_k)]$, respectively. The factors of $S^\lambda \otimes S^{(\mu_1)} \otimes \ldots \otimes S^{(\mu_k)} \uparrow \mathfrak{S}_n$ are given by $[\lambda]^{[\mu_1]^\cdot[\mu_2]^\cdot\ldots[\mu_k]^\cdot}$.

(By $(\lambda,\mu_1,\ldots,\mu_k)$ we mean the partition $(\lambda_1,\ldots,\lambda_j, \mu_1,\ldots,\mu_k)$, where λ_j is the last non-zero part of λ).

Proof: It is simple to see that

$$S^{\lambda,(\lambda,\mu_1,\ldots,\mu_k)} \cong S^\lambda \otimes S^{(\mu_1)} \otimes \ldots \otimes S^{(\mu_k)} \uparrow \mathfrak{S}_n$$

and we just apply Theorem 17.13(ii) to obtain a Specht series. The last sentence is true because $[0]^{[\lambda,(\lambda,\mu_1,\ldots,\mu_k)]^\cdot} = [\lambda]^{[\mu_1]^\cdot\ldots[\mu_k]^\cdot}$ as can be easily verified.

Remark James and Peel have recently constructed a Specht series for the module $S^\mu \otimes S^\lambda \uparrow \mathfrak{S}_n$. Here again, the factors and their order of appearance are independent of the ground field. The Specht factors are given by the Littlewood-Richardson Rule.

17.15 EXAMPLE We construct a Specht series for $M^{(3,2,1)}$. In the tree below, we always absorb the first part of μ into $\mu^\#$ (e.g. $M^{(3,2,1)} = S^{0,(3,2,1)} = S^{(3),(3,2,1)}$; cf. 17.5). Above each picture we give the dimension of the corresponding module.

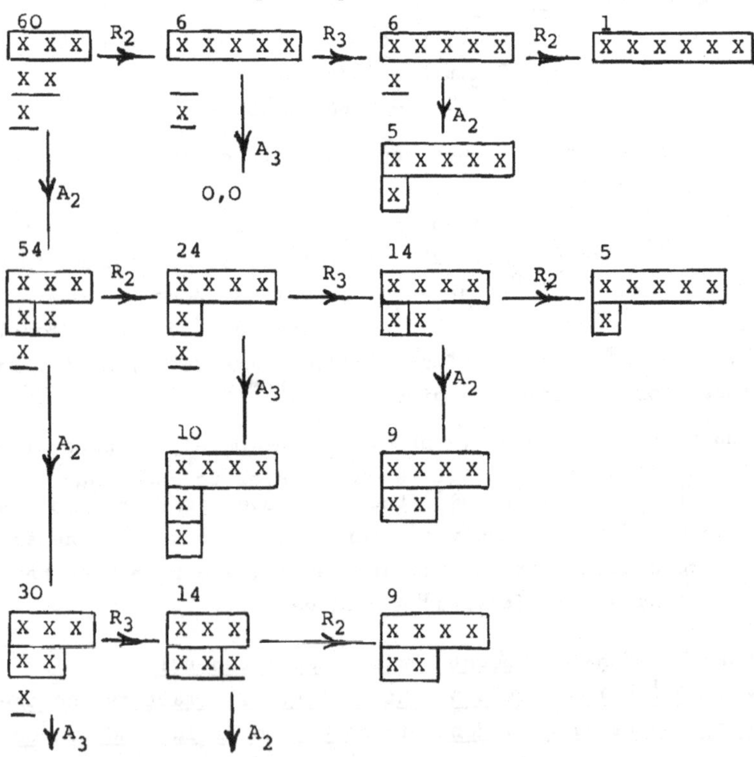

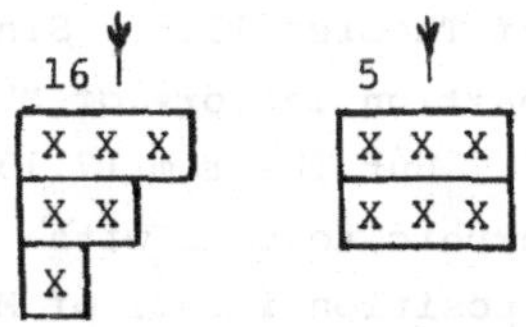

Therefore, $M^{(3,2,1)}$ has a Specht series with factors $S^{(6)}$, $S^{(5,1)}$, $S^{(5,1)}$, $S^{(4,2)}$, $S^{(4,1^2)}$, $S^{(4,2)}$, $S^{(3^2)}$, $S^{(3,2,1)}$, reading from the top. This holds <u>regardless of the ground field</u>.

17.16 EXAMPLE Consider $S^{(4,2^2,1)}\uparrow \mathfrak{S}_{10} = S^{(4,2^2,1),(4,2^2,1^2)}$

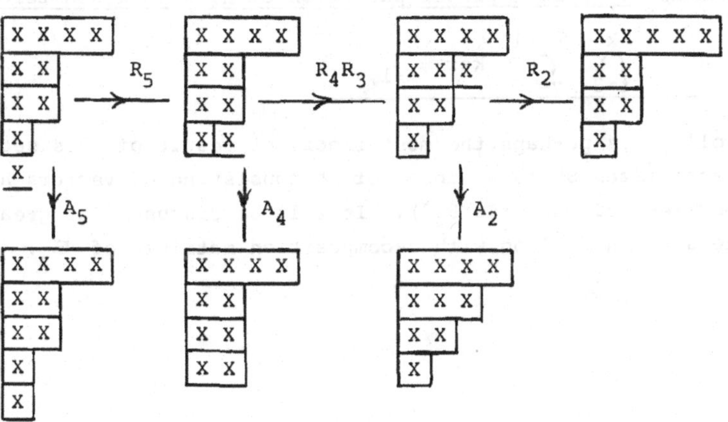

Hence, $S^{(4,2^2,1)}\uparrow \mathfrak{S}_{10}$ has a series with factors, reading from the top, isomorphic to $S^{(5,2^2,1)}$, $S^{(4,3,2,1)}$, $S^{(4,2^3)}$, $S^{(4,2^2,1^2)}$ (cf. Examples 9.1 and 9.5).

17.17 EXAMPLE Following our algorithm, we find that when $m \leq n-m$, $M^{(n-m,m)}$ has a Specht series with factors $S^{(n)}$, $S^{(n-1,1)}$, ..., $S^{(n-m,m)}$, reading from the top (cf. Example 14.4).

There is a point to beware of here. It seems plausible that $M^{(n-m-1,m+1)}/S^{(n-m-1,m+1)}$ is isomorphic to $M^{(n-m,m)}$; after all, both modules have Specht series with factors as listed above. However, this is sometimes false. For instance, when char F = 2, $S^{(6,2)}$ has composition factors $D^{(6,2)}$ and $D^{(7,1)}$ (see the decomposition matrices in the Appendix.) Since $D^{(6,2)}$ is at the top of $S^{(6,2)}$

$$D^{(7,1)} \cong S^{(6,2)} \cap S^{(6,2)\perp} \cong M^{(6,2)}/(S^{(6,2)} + S^{(6,2)\perp}).$$

Therefore, $M^{(6,2)}/S^{(6,2)}$ has a top factor isomorphic to $D^{(7,1)}$, while $M^{(7,1)}$ does not (see Example 5.1).

Theorem 17.13 provides an alternative method of showing that all the irreducible representations of $\mathfrak{S}_n$ appear as a D^ν, thereby avoiding the quotes from Curtis and Reiner in the proof of Theorem 11.5. Since $S^{\mu\perp}$ has the same factors as M^μ/S^μ, all the composition factors of M^μ come from D^μ (if μ is p-regular), and from M^μ/S^μ. But Theorem 17.13 shows that M^μ/S^μ has a series with factors isomorphic to S^λ's with $\lambda \rhd \mu$. By induction, since $S^\lambda \subseteq M^\lambda$, every composition factor of M^μ is isomorphic to some D^ν. Applying this fact to the case where $\mu = (1^n)$, when M^μ is the regular representation of $F\mathfrak{S}_n$, Theorem 1.1 shows that every irreducible $F\mathfrak{S}_n$-module is isomorphic to some D^ν.

Theorem 17.13(i) has the useful

17.18 COROLLARY <u>If μ is a proper partition of n, with k non-zero parts, then</u>

$$S^\mu = \bigcap_{i=2}^{k} \bigcap_{v=0}^{\mu_i - 1} \ker \psi_{i-1,v} .$$

The Corollary is perhaps the most important result of this section, since it characterizes S^μ as a subset of M^μ consisting of vectors having certain properties (cf. Example 5.2). It will be discussed at greater length in the section dealing with decomposition matrices of $\mathfrak{S}_n$.

18 HOOKS AND SKEW-HOOKS

Hooks play an important part in the representation theory of $\mathfrak{S}_n$, but it is not clear in terms of modules why they have a rôle at all! For example, it would be nice to have a direct proof of the Hook formula for dimensions (section 20), without doing all the work required for the standard basis of the Specht module.

The (i,j)-hook may be regarded as the intersection of an infinite Γ shape (having the (i,j)-node at its corner) with the diagram.

18.1 EXAMPLE

```
X X X X        The (2,2)-hook is   X X X X
X X X X                            X ⊠ ⊠ ⊠
X X X                              X ⊠ X
```

and the hook graph is

```
6 5 4 2
5 4 3 1
3 2 1
```

18.2 DEFINITIONS

(i) The <u>(i,j)-hook</u> of $[\mu]$ consists of the (i,j)-node along with the $\mu_i - j$ nodes to the right of it (called the <u>arm</u> of the hook) and the $\mu'_j - i$ nodes below it (called the <u>leg</u> of the hook).

(ii) The <u>length</u> of the (i,j)-hook is $h_{ij} = \mu_i + \mu'_j + 1 - i - j$

(iii) If we replace the (i,j)-node of $[\mu]$ by the number h_{ij} for each node, we obtain the <u>hook graph</u>.

(iv) A <u>skew-hook</u> is a connected part of the rim of $[\mu]$ which can be removed to leave a proper diagram.

18.3 EXAMPLE

```
X X X X      and      X X X X      show the only two
X X X-X               X X X-X
X X-X                 X X X
```

skew 4-hooks in $[4^2,3]$. The diagram also has one skew 6-hook, two skew 5-hooks, two skew 3-hooks, two skew 2-hooks, and two skew 1-hooks. Comparing this with the hook graph, we have illustrated:

18.4 LEMMA <u>There is a natural 1-1 correspondence between the hooks of $[\mu]$ and the skew-hooks of $[\mu]$.</u>

<u>Proof</u>: The skew hook

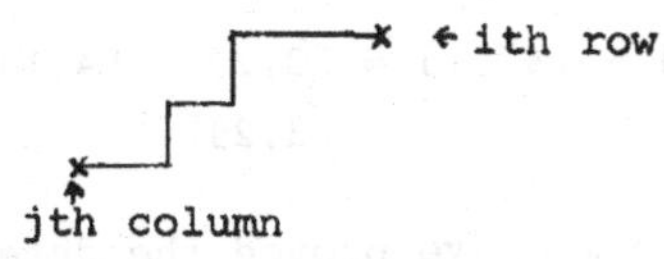

corresponds to the (i,j)-hook.

19 THE DETERMINANTAL FORM

We have seen that when $\lambda_1 \geq \lambda_2 \geq \ldots.,$

$$[\lambda_1][\lambda_2][\lambda_3].. = \sum_\mu m_{\lambda\mu} [\mu]$$

and the matrix $m = (m_{\lambda\mu})$ is lower triangular with 1's down the diagonal (see 6.4 and 4.13). It follows that

$$[\lambda] = \sum_\mu (m^{-1})_{\lambda\mu} [\mu_1][\mu_2][\mu_3]....$$

and m^{-1} is lower triangular with 1's down the diagonal.

19.1 EXAMPLE Inverting the matrix m for $\mathfrak{S}_5$ given in section 6, we find

$m^{-1} =$

	[5]	[4][1]	[3][2]	[3][1]^2	[2]^2[1]	[2][1]^3	[1]^5
[5]	1						
[4,1]	-1	1					
[3,2]	0	-1	1				
[3,1^2]	1	-1	-1	1			
[2^2,1]	0	1	-1	-1	1		
[2,1^3]	-1	1	2	-1	-2	1	
[1^5]	1	-2	-2	3	3	-4	1

The coefficients in the matrix m are given by Young's Rule, and the entries in m^{-1} can be found directly by

19.2 THE DETERMINANTAL FORM <u>If λ is a proper partition of n, then</u>

$$[\lambda] = |[\lambda_i - i+j]|$$

<u>where we define [m] = 0 if m < 0.</u>

The way to write down the determinant for $[\lambda]$ is to put $[\lambda_1],[\lambda_2]..$ in order down the diagonal, and then let the numbers increase by 1 as we go from one term to the next in each row. Beware of the distinction between [0] (which behaves as a multiplicative identity) and 0 (0 × anything = 0).

19.3 EXAMPLES

$$\begin{vmatrix} [3] & [4] \\ [0] & [1] \end{vmatrix} = [3][1] - [4] = [3,1] + [4] - [4] = [3,1]$$

$$\begin{vmatrix} [3] & [4] \\ [1] & [2] \end{vmatrix} = [3][2] - [4][1] = [3,2] + [4,1] + [5] - [4,1] - [5]$$
$$= [3,2]$$

19.4 EXAMPLE Suppose we have proved the determinantal form for 2-

part partitions. Then expanding the following determinant up the last column, we have

$$\begin{vmatrix} [3] & [4] & [5] \\ [1] & [2] & [3] \\ [0] & [1] & [2] \end{vmatrix} = \begin{vmatrix} [3] & [4] \\ [1] & [2] \end{vmatrix} [2] - \begin{vmatrix} [3] & [4] \\ [0] & [1] \end{vmatrix} [3] + \begin{vmatrix} [1] & [2] \\ [0] & [1] \end{vmatrix} [5]$$

which, by induction, is $[3,2][2] - [3,1][3] + [1^2][5]$

$$\begin{aligned} = \; & [3,2^2] + [3^2,1] + [4,2,1] + [4,3] + [5,2] \\ & -([3^2,1] + [4,2,1] + [4,3] + [5,2]) - ([6,1] + [5,1^2]) \\ & + [6,1] + [5,1^2] = [3,2^2] \end{aligned}$$

↑ Diagrams containing

X X ⊠
X ⊠
X ⊠

↑ Diagrams containing

X X ⊠
X ⊠

↑ Diagrams containing

X X ⊠

<u>Proof of the Determinantal Form</u>: It is sufficient to prove the result in the case where $\lambda = (\lambda_1,\lambda_2,\ldots,\lambda_k)$ with $\lambda_k > 0$, since zero parts at the end of λ do not change the determinant. The result is true when λ has no non-zero part, so assume that we have proved the result for λ having fewer than k non-zero parts.

The numbers in the last column of $([\lambda_i - i+j])$ are the "first column hook lengths of $[\lambda]$", $h_{11},h_{21},\ldots,h_{k1}$, since $h_{i1} = \lambda_i + \lambda_1' + 1 - i - 1 = \lambda_i - i + k$.

Let s_i be the skew hook of $[\lambda]$ corresponding to the $(i,1)$-hook (In Example 19.4, s_3, s_2 and s_1 are

X X X
X X
⊠ ⊠

X X X
X ⊠
⊠ ⊠

X ⊠ ⊠
X ⊠
⊠ ⊠

).

Omitting the last column and ith row of $([\lambda_i - i+j])$ gives a matrix with diagonal terms

$$[\lambda_1],[\lambda_2],\ldots,[\lambda_{i-1}],[\lambda_{i+1} - 1],\ldots,[\lambda_k - 1]$$

and these are precisely the parts of $[\lambda \setminus s_i]$. Therefore, the result of expanding the determinant $|[\lambda_i - i+j]|$ up the last column and using induction is

$$[\lambda\setminus s_k][h_{k1}] - [\lambda\setminus s_{k-1}][h_{k-1,1}]+\ldots\pm[\lambda\setminus s_1][h_{11}] \qquad (*)$$

Now consider $[\lambda \setminus s_i][h_{i1}]$. This is evaluated by adding h_{i1} nodes to $[\lambda \setminus s_i]$ in all ways such that no two added nodes are in the same column (by the Littlewood-Richardson Rule, or Corollary 17.14). $[\lambda \setminus s_i]$ certainly contains the last node of the 1st, 2nd,...,(i-1)th rows of $[\lambda]$, so we deduce that all the diagrams in $[\lambda \setminus s_i][h_{i1}]$

(i) contain the last nodes of the 1st,2nd,...,(i-1)th rows of $[\lambda]$, and (ii) do not contain the last nodes of the (i+1)th, (i+2)th,...,kth rows of $[\lambda]$.

Split the diagrams in $[\lambda \setminus s_i][h_{i1}]$ into 2 set, according to whether or not the last node of the ith row of $[\lambda]$ is in the diagram. It is clear that $[\lambda]$ is the only diagram we get containing the last nodes of all the rows of $[\lambda]$, and a little thought shows that in (*) we get sets cancelling in pairs to leave $[\lambda]$. This proves the Determinantal Form.

19.5 COROLLARY <u>$\dim S^\lambda = n! \left| \frac{1}{(\lambda_i - i+j)!} \right|$ where $\frac{1}{r!} = 0$ if $r < 0$</u>

<u>Proof</u>: $[\mu_1][\mu_2]\ldots[\mu_k]$ has dimension $\dfrac{n!}{\mu_1!\ldots\mu_k!}$ (see 4.2), and the Corollary is now immediate.

20 THE HOOK FORMULA FOR DIMENSIONS

20.1 THEOREM (Frame, Robinson and Thrall [4])

The dimension of the Specht module S^λ is given by

$$\dim S^\lambda = n! \frac{\prod_{i<k} (h_{i1} - h_{k1})}{\prod_i h_{i1}!} = \frac{n!}{\prod(\text{hook lengths in } [\lambda])}$$

20.2 EXAMPLE The hook graph for [4,3,1] is

6 4 3 1
4 2 1
1

Therefore, $\dim S^{(4,3,1)} = \frac{8!}{6.4.3.4.2} = 70.$

The hook formula is an amazing result. It is hard to prove directly even that n! is divisible by the product of the hook lengths, let alone show that the quotient is the number of standard λ-tableaux.

Proof of Theorem 20.1 We show that the result is true when λ has 3 non-zero parts. It is transparent that the proof works in general, but a full proof obscures the simplicity of the ideas required.

By Corollary 19.5,

$$\frac{\dim S^\lambda}{n!} = \begin{vmatrix} \frac{1}{(h_{11}-2)!} & \frac{1}{(h_{11}-1)!} & \frac{1}{h_{11}!} \\ \frac{1}{(h_{21}-2)!} & \frac{1}{(h_{21}-1)!} & \frac{1}{h_{21}!} \\ \frac{1}{(h_{31}-2)!} & \frac{1}{(h_{31}-1)!} & \frac{1}{h_{31}!} \end{vmatrix}$$

$$= \frac{1}{h_{11}!} \frac{1}{h_{21}!} \frac{1}{h_{31}!} \begin{vmatrix} h_{11}(h_{11}-1) & h_{11} & 1 \\ h_{21}(h_{21}-1) & h_{21} & 1 \\ h_{31}(h_{31}-1) & h_{31} & 1 \end{vmatrix}$$

$$= \frac{(h_{11}-h_{21})(h_{11}-h_{31})(h_{21}-h_{31})}{h_{11}!\, h_{21}!\, h_{31}!}$$ giving the first result.

$$= \frac{1}{h_{11}!}\ \frac{1}{h_{21}!}\ \frac{1}{h_{31}!} \begin{vmatrix} (h_{11}-1)(h_{11}-2) & h_{11}-1 & 1 \\ (h_{21}-1)(h_{21}-2) & h_{21}-1 & 1 \\ (h_{31}-1)(h_{31}-2) & h_{31}-1 & 1 \end{vmatrix}$$

$$= \frac{1}{h_{11}\ h_{21}\ h_{31}} \begin{vmatrix} \frac{1}{(h_{11}-3)!} & \frac{1}{(h_{11}-2)!} & \frac{1}{(h_{11}-1)!} \\ \frac{1}{(h_{21}-3)!} & \frac{1}{(h_{21}-2)!} & \frac{1}{(h_{21}-1)!} \\ \frac{1}{(h_{31}-3)!} & \frac{1}{(h_{31}-2)!} & \frac{1}{(h_{31}-1)!} \end{vmatrix}$$

$$= \frac{1}{h_{11}\ h_{21}\ h_{31}} \quad \frac{1}{\Pi(\text{hook lengths in } [\lambda_1-1,\lambda_2-1,\lambda_3-1])}\,,$$

by induction

$$= \frac{1}{\Pi(\text{hook lengths in } [\lambda])}\,, \text{ as required.}$$

21 THE MURNAGHAN-NAKAYAMA RULE

The Murnaghan-Nakayama Rule is a very beautiful and efficient way of calculating a single entry in the character table of $\mathfrak{S}_n$.

In the statement below, the leg-length of a skew-hook is defined to be the same as that of the corresponding hook.

21.1 THE MURNAGHAN-NAKAYAMA RULE

<u>Suppose that $\pi\rho \in \mathfrak{S}_n$ where ρ is an r-cycle and π is a permutation of the remaining n-r numbers. Then</u>

$$\chi^{\lambda}(\pi\rho) = \sum_{\nu} \{(-1)^i \chi^{\nu}(\pi) \mid [\lambda] \setminus [\nu] \text{ is a skew r-hook of leg length } i\}.$$

As usual, an empty sum is interpreted as zero. The case where ρ is a 1-cycle is the Branching Theorem.

21.2 EXAMPLES

(i) Suppose we want to find the value of $\chi^{(5,4,4)}$ on the class $(5,4,3,1)$.

```
X X X X X      X X X X X      X X X X X
X X X X        X X X X        X X X X
X X X X        X X X X        X X X X
```

There are two ways of removing a skew 5-hook from $[5,4,4]$ and the Murnaghan-Nakayama Rule gives:

$$\chi^{(5,4,4)} \text{ on } (5,4,3,1) = \chi^{(3,3,2)} - \chi^{(5,3)} \text{ on } (4,3,1)$$

$$= \chi^{(2,1^2)} - \chi^{(3,1)} + \chi^{(2^2)} \text{ on } (3,1),$$

applying the rule again

$$= \chi^{(2^2)} \text{ on } (3,1),$$ because we cannot remove a skew 3-hook from either $[2,1^2]$ or $[3,1]$.

$$= -\chi^{(1)} \text{ on } (1)$$

$$= -1.$$

(ii) $\chi^{(5,4,4)}$ is zero on any class containing an 8,9,10,11,12 or 13-cycle, since we cannot remove hooks of these lengths from $[5,4,4]$.

(iii)

```
X X X X X      X X X X X
X X X X        X X X X
X X X X        X X X X
```

$$\chi^{(5,4,4)} \text{ on } (7,3,3) = \chi^{(3^2)} \text{ on } (3,3)$$

$$= -\chi^{(2,1)} + \chi^{(3)} \text{ on } (3)$$

$$= \chi^{(0)} + \chi^{(0)} \text{ on } (0)$$

$$= 2.$$

The only character table required in the construction of the character table of $\mathfrak{S}_n$ using the Murnaghan-Nakayama Rule is that of $\mathfrak{S}_0$. Remember that $\mathfrak{S}_0$ is a group of order 1, and a computer is unnecessary in evaluating the character table of $\mathfrak{S}_0$!

Our proof of the Murnaghan-Nakayama Rule needs several preliminary lemmas. We first prove the special case where ρ is an n-cycle, then examine what the Littlewood-Richardson Rule gives for $[\nu][x,1^{r-x}]$, and finally we combine these pieces of information to prove the Rule in general. See the remarks following 21.12 for an alternative approach.

A hook diagram is one of the form $[x,1^y]$.

21.3 LEMMA <u>Unless both $[\alpha]$ and $[\beta]$ are hook diagrams, $[\alpha][\beta]$ contains no hook diagrams. If $[\alpha] = [a,1^{n-r-a}]$ and $[\beta] = [b,1^{r-b}]$ then $[\alpha][\beta] = [a+b,1^{n-a-b}] + [a+b-1,1^{n-a-b+1}]$ + some non-hook diagrams.</u>

<u>Proof</u>: If one of $[\alpha]$ and $[\beta]$ contains the (2,2)-node, the so does $[\alpha][\beta] = [\alpha]^{[\beta]^{\bullet}} = [\beta]^{[\alpha]^{\bullet}}$. This proves the first result.

Suppose, therefore, that $[\alpha] = [a,1^{n-r-a}]$ and $[\beta] = [b,1^{r-b}]$. In order to obtain a hook diagram in $[\alpha]^{[\beta]^{\bullet}}$, we have to put b 1's in the places shown, then 2,3,... in order down the first column:

```
                      b
                  ___/\___
          X X.....X  * * ... *
          X
[α] →     .
          .
          X
          *
```

The second result follows.

21.4 THEOREM (A special case of the Murnaghan-Nakayama Rule). <u>Suppose that ρ is an n-cycle, and ν is a partition of n. Then</u>

$$\chi^{\nu}(\rho) = \begin{cases} (-1)^{n-x} & \underline{\text{if}}\ [\nu] = [x,1^{n-x}] \\ 0 & \underline{\text{otherwise}} \end{cases}$$

<u>Proof</u>: Let $[\alpha]$ and $[\beta]$ be diagrams for $\mathfrak{S}_r$ and $\mathfrak{S}_{n-r}$ with $0 < r < n$. Then the character inner product

$$(\chi^{[\alpha][\beta]}, \chi^{(n)-(n-1,1)+(n-2,1^2)- \ldots \pm (1^n)})$$

is zero, since $[\alpha][\beta]$ contains two adjacent hook diagrams, each with coefficient 1, or no hooks at all by Lemma 21.3.

By the Frobenius Reciprocity Theorem, $\chi^{(n)-(n-1,1)+\ldots\pm(1^n)}$ restricts to be zero on all Young subgroups of the form $\mathfrak{S}_{(r,n-r)}$ with $0 < r < n$; in particular, it has value zero on all classes of $\mathfrak{S}_n$, except perhaps, that containing our n-cycle ρ. Therefore, the column vector which has $(-1)^{n-x}$ opposite $\chi^{(x,1^{n-x})}$ and 0 opposite all other characters is orthogonal to all columns of the character table of $\mathfrak{S}_n$, except that associated with ρ. Since the character table is non-singular, this column vector must be a multiple of the ρ-column. But the entry opposite $\chi^{(n)}$ is 1. Therefore, it is the ρ-column, as required.

Remark: Theorem 21.4 can also be proved using the Determinantal Form, but the above proof is more elegant.

21.5 LEMMA Suppose that λ is a partition of n and ν is a partition of n-r. Then

(i) The multiplicity of $[\lambda]$ in $[\nu][x,1^{r-x}]$ is zero unless $[\lambda]\setminus[\nu]$ is a union of skew-hooks.

(ii) The multiplicity of $[\lambda]$ in $[\nu][x,1^{r-x}]$ is the binomial coefficient $\binom{m-1}{c-x}$ if $[\lambda]\setminus[\nu]$ is a union of m disjoint skew hooks having (in total) c columns (and r nodes).

Proof: The Littlewood-Richardson Rule assures us that the diagram $[\lambda]$ appears in $[\nu][x,1^{r-x}]$ if and only if $[\nu]$ is a subdiagram of $[\lambda]$ and we can replace the nodes in $[\lambda]\setminus[\nu]$ by x 1's, one 2, one 3,..., one (r-x) in such a way that

(i) Any column containing a 1 has just one 1, which is at the top of the column.

(ii) For $i > 1$, i+1 is in a later row than i; in particular, no two numbers greater than 1 are in the same row.

(iii) The first non-empty row contains no number greater than 1.

(iv) Any row containing a number greater than 1 has it at the end of the row.

Suppose that the multiplicity of $[\lambda]$ in $[\nu][x,1^{r-x}]$ is non-zero. Then $[\lambda]\setminus[\nu]$ does not contain four nodes in the shape

X X
X X

since neither left hand node can be replaced by a number greater than 1 (by (iv)); nor can they both be replaced by 1 (by (i)). Therefore, $[\lambda]\setminus[\nu]$ is a union of skew hooks.

Suppose that $[\lambda]\setminus[\nu]$ is a union of m disjoint skew-hooks, having c columns. When we try to replace the nodes in $[\lambda]\setminus[\nu]$ by numbers, we notice that certain nodes must be replaced by 1's and others by some numbers $b > 1$, as in the following example

```
                          1 1
                          b
                      1 1 X
                  1 1 b
                  b
          1 1 X                      c = 11, m = 4
          b
      X
      b
      b
```

Each column contains at most one 1 (by (i)). Also, each column contains at least one 1, except the last column of the 2nd, 3rd,..., mth components (by (ii),(iii) and (iv)). Therefore, (c-m+1) 1's are forced. There remain (x-c + m-1) 1's which can be put in any of the m-1 spaces left at the top of the last columns in the 2nd, 3rd,...,mth components. The position of each number greater than 1 is determined by (ii) once the 1's have been put in. The multiplicity of $[\lambda]$ in $[\nu][x,1^{r-x}]$ is therefore $\binom{m-1}{x-c+m-1} = \binom{m-1}{c-x}$, as we claimed.

<u>Proof of the Murnaghan-Nakayama Rule</u>:

Let $a_{\nu\mu} = (\chi^{\lambda}\downarrow \mathfrak{S}_{(n-r,r)},\ \chi^{[\nu][\mu]})$, where μ is a partition of r and ν is a partition of n-r.

If ρ is an r-cycle and π is a permutation of the remaining n-r numbers, then

$$\chi^{\lambda}(\pi\rho) = \sum_{\nu,\mu} a_{\nu\mu}\chi^{\nu}(\pi)\ \chi^{\mu}(\rho)$$

$$= \sum_{\nu} \chi^{\nu}(\pi) \sum_{x=1}^{r} a_{\nu,(x,1^{r-x})}(-1)^{r-x}, \quad \text{by 21.4.}$$

But $a_{\nu,(x,1^{r-x})} = (\chi^{\lambda},\ \chi^{[\nu][x,1^{r-x}]})$ by the Frobenius Reciprocity Theorem

$$= \binom{m-1}{c-x} \quad \text{by Lemma 21.5.}$$

The definitions of m and c give $r \geq c \geq m$, so

$$\sum_{x=1}^{r} \binom{m-1}{c-x}(-1)^{r-x} = (-1)^{r-c}\left\{\binom{m-1}{0} - \binom{m-1}{1} + \dots \pm \binom{m-1}{m-1}\right\}$$

$$= \begin{cases} (-1)^{r-c} & \text{if } m = 1 \\ 0 & \text{if } m \neq 1. \end{cases}$$

However, when $m = 1$, $[\lambda] \setminus [\nu]$ is a single skew r-hook of leg length r-c. Therefore,

$$\chi^\lambda(\pi\rho) = \sum_\nu \{(-1)^i \chi^\nu(\pi) \mid [\lambda] \setminus [\nu] \text{ is a skew r-hook of leg length } i\},$$

which is the Murnaghan-Nakayama Rule.

21.6 COROLLARY <u>Suppose p is a prime. If no entry in the hook graph for $[\lambda]$ is divisible by p, then χ^λ is zero on all permutations whose order is divisible by p.</u>

<u>Proof</u>: The hypothesis shows that no skew kp-hook can be removed from $[\lambda]$, so the Murnaghan-Nakayama Rule shows that χ^λ is zero on all permutations containing a kp-cycle $(k > 0)$.

<u>Remark</u> The hypothesis of Corollary 21.6 is equivalent to the statement that $|\mathfrak{S}_n| / \deg \chi^\lambda$ is coprime to p, by the Hook Formula. The Corollary therefore illustrates the general theorem that if χ is an ordinary irreducible character of a group G and $|G| / \deg \chi$ is coprime to p, then χ is zero on all p-singular elements of G. (In the language of modular theory, χ is in a <u>block of defect 0</u>.)

The Murnaghan-Nakayama Rule can be rephrased in a way which is useful in numerical calculations, especially in the modular theory for $\mathfrak{S}_n$.

21.7 THEOREM <u>If ν is a partition of n-r, then the generalised character of $\mathfrak{S}_n$ corresponding to</u>

$$\sum \{(-1)^i[\lambda] \mid [\lambda] \setminus [\nu] \text{ is a skew r-hook of leg-length } i\}$$

<u>is zero on all classes except those containing an r-cycle.</u>

<u>Proof</u>: Suppose that $[\lambda]$ is a diagram appearing in

$$[\nu]([r] - [r-1,1] + [r-2,1^2] - \dots \pm [1^r]).$$

Then, by Lemma 21.5, $[\lambda] \setminus [\nu]$ is a union of m disjoint skew hooks and its coefficient is

$$\sum_{x=1}^{r} \binom{m-1}{c-x} (-1)^{r-x}$$

As before, this is $(-1)^{r-c}$ if $m = 1$, and zero if $m \neq 1$. Therefore

$$[\nu]([r] - [r-1,1] + [r-2,1^2] - \dots \pm [1^r])$$
$$= \sum \{(-1)^i[\lambda] \mid [\lambda] \setminus [\nu] \text{ is a skew r-hook of leg length } i\}.$$

But, by definition, $\chi^\nu \chi^{(r)-(r-1,1)+ \dots \pm (1^r)} \uparrow \mathfrak{S}_n$ is zero on all of $\mathfrak{S}_n$ except the subgroup $\mathfrak{S}_{(n-r,r)}$. However, it is zero even here, except on $\pi\rho$ (ρ an r-cycle), by Theorem 21.4.

Remark: The proof shows that "the operator $[r]^\bullet - [r-1,1]^\bullet + \ldots \pm [1^r]^\bullet$ wraps skew r-hooks on to the rim of a diagram".

21.8 EXAMPLES (i) When $\nu = (3,2)$ and $r = 3$

```
+ X X X . . .    - X X X .    - X X X     + X X X
  X X              X X . .      X X         X X
                                . .         .
                                .           .
                                            .
```

shows the ways of wrapping skew 3-hooks on to [3,2]. The generalised character $\chi^{(6,2)} - \chi^{(4^2)} - \chi^{(3,2^2,1)} + \chi^{(3,2,1^3)}$ is zero on all classes of $\mathfrak{S}_8$ except those containing a 3-cycle.

(ii) For $n \geq 4$, $\chi^{(n)} + \chi^{(n-2,2)} - \chi^{(n-2,1^2)}$ is zero on all classes of $\mathfrak{S}_n$ except those containing a 2-cycle.

These examples show that $\chi^{(6,2)} + \chi^{(3,2,1^3)} = \chi^{(4^2)} + \chi^{(3,2^2,1)}$ as a 3-modular character, since this equation holds on 3-regular classes, and $\chi^{(n-2,1^2)} = \chi^{(n-2,2)} + \chi^{(n)}$ as a 2-modular character. At once, it follows that $\chi^{(n-2,1^2)}$, $\chi^{(n-2,2)}$ and $\chi^{(n)}$ are in the same 2-block of $\mathfrak{S}_n$. Also, $\chi^{(6,2)}$, $\chi^{(3,2,1^3)}$, $\chi^{(4^2)}$ and $\chi^{(3,2^2,1)}$ are in the same 3-block of $\mathfrak{S}_8$, since

21.9 THEOREM <u>Let $\Sigma a_\lambda \chi^\lambda = 0$ be a non-trivial relation between characters on p-regular classes. Then a_λ is non-zero for some p-singular λ, and if a_λ is non-zero for just one p-singular λ, then all the characters with non-zero coefficients are in the same p-block.</u>

<u>Proof</u>: If the only non-zero coefficients belong to p-regular partitions, consider the last partition μ whose coefficient a_μ is non-zero. The character χ^μ contains a modular irreducible character ϕ^μ corresponding to the factor D^μ of S^μ. By Corollary 12.2, ϕ^μ is not a constituent of any other ordinary character in our relation, and this contradicts the fact that the modular irreducible characters of a group are linearly independent.

If the partitions with non-zero coefficients lie in more than one p-block, then there are two non-trivial subrelations of the given one, and each subrelation must involve a p-singular partition, by what we have just proved. The Theorem now follows.

Although it is fairly easy to prove that all relations between the ordinary characters of $\mathfrak{S}_n$, regarded as p-modular characters, come from applying Theorem 21.7, there seems to be no way of completely determining the p-block structure of $\mathfrak{S}_n$ along these lines.

21.10 EXAMPLE It is an easy exercise to prove from the Murnaghan-Nakayama Rule that when $n = 2m$ is even

$$\chi^{(n)} - \chi^{(n-1,1)} + \chi^{(n-2,2)} - \dots \pm \chi^{(m,m)}$$

is zero on all classes of $\mathfrak{S}_n$ containing an odd cycle. Hence $\chi^{(n)}, \chi^{(n-1,1)}, \dots, \chi^{(m,m)}$ are all in the same 2-block of $\mathfrak{S}_{2m}$, by Theorem 21.9.

This is a convenient point at which to state

21.11 THEOREM ("The Nakayama Conjecture"). <u>S^μ and S^λ are in the same p-block of $\mathfrak{S}_n$ if and only if there is a (finite) permutation σ of $\{1,2,\dots\}$ such that for all i</u>

$$\lambda_i - i \equiv \mu_{i\sigma} - i\sigma \quad \text{modulo } p.$$

We do not prove the Nakayama Conjecture here - the interested reader is referred to Meier and Tappe [17] where the latest proof and references to all earlier ones appear. It seems to the author that the value of this Theorem has been overrated; it is certainly useful (but not essential) when trying to find the decomposition matrix of $\mathfrak{S}_n$ for a particular small n, but there are few general theorems in which it is helpful. In fact, there is just one case of the Nakayama Conjecture needed for a Theorem in this book, and we prove this now:

21.12 LEMMA <u>If n is odd, $S^{(n)}$ and $S^{(n-1,1)}$ are in different 2-blocks of $\mathfrak{S}_n$.</u>

<u>Proof</u>: Let $\pi = (1\ 2)(3\ 4)\dots(n-2,n-1)$. Then $|\mathcal{C}_\pi|$ is odd, where $\mathcal{C}_\pi$ is the conjugacy class of $\mathfrak{S}_n$ containing π. But $\chi^{(n)}(\pi) = 1$ and $\chi^{(n-1,1)}(\pi) = 0$, by Lemma 6.9. Therefore,

$$|\mathcal{C}_\pi| \frac{\chi^{(n)}(\pi)}{\chi^{(n)}(1)} \not\equiv |\mathcal{C}_\pi| \frac{\chi^{(n-1,1)}(\pi)}{\chi^{(n-1,1)}(1)} \quad \text{mod } 2.$$

General theory (see Curtis and Reiner [2], 85.12) now tells us that $S^{(n)}$ and $S^{(n-1,1)}$ are in different 2-blocks.

The proof we have given for the Murnaghan-Nakayama Rule has been designed to demonstrate the way in which skew-hooks come into play. The Rule can also be deduced from the Determinantal Form, and we conclude this section with an outline of the method.

21.12 LEMMA <u>Suppose that $\pi\rho \in \mathfrak{S}_n$ where ρ is an r-cycle and π is a permutation of the remaining n-r numbers. Let $(\mu_1,\mu_2,\dots,\mu_n)$ be a partition of n. Then</u>

$$\chi^{[\mu_1][\mu_2]\dots[\mu_n]}(\pi\rho) = \sum_{i=1}^{n} \chi^{[\mu_1][\mu_2]\dots[\mu_{i-1}][\mu_i - r][\mu_{i+1}]\dots[\mu_n]}(\pi).$$

<u>Proof:</u> $\chi^{[\mu_1]\ldots[\mu_n]}(\pi\rho)$ = the number of μ-tabloids fixed by $\pi\rho$

$= \sum_{i=1}^{n}$ (the number of μ-tabloids fixed by π in which all the numbers moved by ρ lie in the ith row), since a μ-tabloid is fixed by ρ if and only if each orbit of ρ is contained in a single row of the tabloid.

$= \sum_{i=1}^{n}$ (the number of $(\mu_1,\mu_2,\ldots,\mu_{i-1},\mu_i-r,\mu_{i+1},\ldots,\mu_n)$-tabloids fixed by π)

$= \sum_{i=1}^{n} \chi^{[\mu_1][\mu_2]\ldots[\mu_{i-1}][\mu_i-r][\mu_{i+1}]\ldots[\mu_n]}(\pi)$, as we wished to show.

As usual, [k] is taken to be zero if $k < 0$, and $\chi^0(\pi) = 0$.

21.14 EXAMPLE (cf. Example 21.2(i)). Suppose that $\pi\rho \in \mathfrak{S}_{13}$ where ρ is a 5-cycle and π is a permutation of the remaining 8 numbers. Then

$\chi^{(5,4,4)}(\pi\rho)$ = the character of $\begin{vmatrix} [5] & [6] & [7] \\ [3] & [4] & [5] \\ [2] & [3] & [4] \end{vmatrix}$ evaluated at $\pi\rho$, by the Determinantal Form

$$= \begin{vmatrix} [0] & [1] & [2] \\ [3] & [4] & [5] \\ [2] & [3] & [4] \end{vmatrix} + \begin{vmatrix} [5] & [6] & [7] \\ [-2] & [-1] & [0] \\ [2] & [3] & [4] \end{vmatrix} + \begin{vmatrix} [5] & [6] & [7] \\ [3] & [4] & [5] \\ [-3] & [-2] & [-1] \end{vmatrix} \text{ at } \pi, \text{ by Lemma 21.13}$$

$$= \begin{vmatrix} [3] & [4] & [5] \\ [2] & [3] & [4] \\ [0] & [1] & [2] \end{vmatrix} - \begin{vmatrix} [5] & [6] & [7] \\ [2] & [3] & [4] \\ [-2] & [-1] & [0] \end{vmatrix} \text{ at } \pi$$

$= (\chi^{(3,3,2)} - \chi^{(5,3,0)})(\pi)$, by the Determinantal Form.

By inspecting the above example, the reader will see what is required to prove the Murnaghan-Nakayama Rule from the Determinantal Form, and should have no difficulty with the details.

22 BINOMIAL COEFFICIENTS

In the next couple of sections, we shall put our mind to the representations of $\mathfrak{S}_n$ over a field of finite characteristic p. Many of the problems which arise depend upon deciding whether or not the prime p divides certain binomial coefficients, and the relevant Lemmas are collected in this section.

22.1 DEFINITION Suppose $n = n_0 + n_1 p + \ldots + n_r p^r$ where, for each i, $0 \le n_i < p$ and $n_r \ne 0$. Then let

(i) $\nu_p(n) = \max \{i \mid n_j = 0 \text{ for } j < i\}$

(ii) $\sigma_p(n) = n_0 + n_1 + \ldots + n_r$

(iii) $\ell_p(n) = r + 1.$

For a positive rational number n/m, let $\nu_p(n/m) = \nu_p(n) - \nu_p(m)$. We do not define $\nu_p(0)$, but we let $\sigma_p(0) = \ell_p(0) = 0$.

22.2 LEMMA $\underline{\nu_p(n!) = (n - \sigma_p(n))/(p - 1)}$.

Proof: The result is true for n = 0, so we may apply induction. If $n = p^r$, then $\nu_p\{(p^r-1)!\} = (p^r-1-rp+r)/(p-1)$, by induction. But $\nu_p(p^r!) = r+\nu_p\{(p^r-1)!\} = (p^r-1)/(p-1)$, and the result is true in this case. Assume, therefore, that $0 < n-p^r < p^{r+1} - p^r$. Since $\nu_p(p^r+ x) = \nu_p(x)$ for $0 < x < p^{r+1} - p^r$,

$$\nu_p\{n(n-1)\ldots(p^r+ 1)\} = \nu_p\{(n-p^r)!\} .$$

Therefore $\nu_p(n!) = \nu_p(p^r!) + \nu_p\{(n - p^r)!\}$

$$= (p^r - 1 + n - p^r - \sigma_p(n) + 1)/(p-1),$$

by induction, and this is the required result.

22.3 LEMMA $\underline{\text{Assume } a \ge b > 0. \text{ Then } \nu_p\binom{a}{b} < \ell_p(a) - \nu_p(b).}$

Proof: We may apply induction on a, since the result is true for a = 1.

If $p \mid b$, let $b' = b/p$ and $a' = (a-a_0)/p$, where $0 \le a_0 < p$ and $a \equiv a_0$ modulo p. Using the last Lemma, we have

$$\begin{aligned}\nu_p\binom{a}{b} &= \{\sigma_p(b) + \sigma_p(a-b) - \sigma_p(a)\}/(p - 1)\\ &= \{\sigma_p(b') + \sigma_p(a'-b') - \sigma_p(a')\}/(p - 1)\\ &= \nu_p\binom{a'}{b'} .\end{aligned}$$

But $\nu_p\binom{a'}{b'} < \ell_p(a') - \nu_p(b')$, by induction, and $\ell_p(a) = \ell_p(a') + 1$ and $\nu_p(b) = \nu_p(b') + 1$, so $\nu_p\binom{a}{b} < \ell_p(a) - \nu_p(b)$, in this case.

Now suppose that $\nu_p(b) = 0$. Since $\binom{a}{b} = \frac{a+b-1}{b}\binom{a}{b-1}$,

$$\nu_p\binom{a}{b} = \nu_p(a-b+1) + \nu_p\binom{a}{b-1}.$$

Because the result is true for $b = 1$, we may assume that $b > 1$, and $\nu_p\binom{a}{b-1} < \ell_p(a) - \nu_p(b-1)$. Hence, unless $\nu_p(a-b+1) > 0$,

$$\nu_p\binom{a}{b} < \ell_p(a).$$

But if $\nu_p(a-b+1) > 0$, then

$$\nu_p\binom{a}{b-1} = \nu_p\binom{a}{a-b+1} < \ell_p(a) - \nu_p(a-b+1),$$

by the first paragraph of the proof. Therefore, $\nu_p\binom{a}{b} < \ell_p(a) = \ell_p(a) - \nu_p(b)$ in this case also.

22.4 LEMMA <u>Assume that</u>

$$a = a_0 + a_1p + \dots + a_rp^r \qquad (0 \le a_i < p)$$

$$b = b_0 + b_1p + \dots + b_rp^r \qquad (0 \le b_i < p).$$

<u>Then</u> $\binom{a}{b} \equiv \binom{a_0}{b_0}\binom{a_1}{b_1} \dots \binom{a_r}{b_r}$ <u>modulo p. In particular, p divides</u> $\binom{a}{b}$ <u>if and only if</u> $a_i < b_i$ <u>for some i.</u>

<u>Proof</u>: As a polynomial over the field of p elements, we have

$$(x+1)^a = (x+1)^{a_0}(x^p+1)^{a_1} \dots (x^{p^r}+1)^{a_r}.$$

Comparing coefficients of x^b, we obtain the result.

22.5 COROLLARY <u>Assume</u> $a \ge b \ge 1$. <u>Then all the binomial coefficients</u> $\binom{a}{b}, \binom{a-1}{b-1}, \dots, \binom{a-b+1}{1}$ <u>are divisible by p if and only if</u>

$$a-b \equiv -1 \bmod p^{\ell_p(b)}.$$

<u>Proof</u>: By considering Pascal's Triangle, p divides all the given binomial coefficients if and only if p divides each of

$$\binom{a-b+1}{1}, \binom{a-b+1}{2}, \dots, \binom{a-b+1}{b}.$$

Then the last sentence of the Lemma gives our result.

23 SOME IRREDUCIBLE SPECHT MODULES

The Specht module S^μ is irreducible over fields of characteristic zero, and since every field is a splitting field for $\mathfrak{S}_n$, S^μ is irreducible over field of prime characteristic p if and only if it is irreducible when the ground field has p elements. This then, is the case we shall investigate and, except where otherwise stated, F is the field of order p in this section. The complete classification of irreducible Specht modules is still an open problem, but we tackle special cases below.

23.1 LEMMA <u>Suppose that $\text{Hom}_{F\mathfrak{S}_n}(S^\mu,S^\mu) \cong F$. Then S^μ is irreducible if and only if S^μ is self dual.</u>

<u>Proof</u>: If S^μ is irreducible, then it is certainly self-dual (since its modular character is real.)

Let U be an irreducible submodule of S^μ. If S^μ is self-dual, then there is a submodule V of S^μ with $S^\mu/V \cong U$. Since

$$S^\mu \underset{\text{canon}}{\rightarrow} S^\mu/V \underset{\text{iso}}{\rightarrow} U$$

gives a non-zero element of $\text{Hom}_{F\mathfrak{S}_n}(S^\mu,S^\mu)$, we must have $U = S^\mu$, so S^μ is irreducible.

The hypothesis $\text{Hom}_{F\mathfrak{S}_n}(S^\mu,S^\mu) \cong F$ cannot be omitted from this Lemma (see Example 23.10(iii) below), but Corollary 13.17 shows that the hypothesis holds for most Specht modules.

Before applying the Lemma, we want a result about the integer $g^{\mu'}$ defined in 10.3 as the greatest common divisor of the integers $\langle e_t, e_{t^*} \rangle'$ where e_t and e_{t^*} are polytabloids in $S_{\mathbb{Q}}^{\mu'}$ (μ' being the partition conjugate to μ, and $\langle\ ,\ \rangle'$ being the bilinear form on $M_{\mathbb{Q}}^{\mu'}$).

Remember that $\kappa_t = \sum_{\pi\in C_t} (\text{sgn}\ \pi)\pi$. Let $\rho_t = \sum_{\pi\in R_t} \pi$.

23.2 LEMMA <u>Let the ground field be $\mathbb{Q}$, and t be a μ-tableau. Then</u>

(i) <u>The greatest common divisor of the coefficients of the tabloids involved in $\{t\}\kappa_t\rho_t$ is $g^{\mu'}$.</u>

<u>and</u> (ii) <u>$\{t\}\kappa_t\rho_t\kappa_t = \Pi$ (hook lengths in $[\mu]$) $\{t\}\kappa_t$.</u>

<u>Proof</u>: (i) By definition, $g^{\mu'} = \text{g.c.d.} \langle e_{t'}, e_{t'}\pi \rangle'$ as the permutation π varies. But

$$\begin{aligned}\text{sgn}\ \pi \langle e_{t'}, e_{t'}\pi \rangle' &= \text{sgn}\ \pi \langle \{t'\}, \{t'\}\kappa_{t'}\pi\kappa_{t'} \rangle \\ &= \Sigma\ \{\text{sgn}\ \pi\ \text{sgn}\ \sigma\ \text{sgn}\ \tau \mid \sigma,\tau \in C_{t'},\ \sigma\pi\tau \in R_{t'}\} \\ &= \Sigma\ \{\text{sgn}\ \omega \mid \tau \in C_{t'},\ \omega\tau^{-1}\pi^{-1} \in C_{t'},\ \omega \in R_{t'}\}\end{aligned}$$

$$= \Sigma \{\operatorname{sgn} \omega \mid \tau \in R_t, \ \omega\tau^{-1}\pi^{-1} \in R_t, \ \omega \in C_t\}$$
$$= < \{t\}, \{t\}\kappa_t\rho_t \pi^{-1} >$$
$$= < \{t\}\pi, \{t\}\kappa_t\rho_t >$$

and result (i) follows.

(ii) Corollary 4.7 shows that $\{t\}\kappa_t\rho_t\kappa_t = c\{t\}\kappa_t$ for some $c \in \mathbb{Q}$. To evaluate c, it is best to consider the group algebra $\mathbb{Q}\mathfrak{S}_n$. (See the remarks at the end of section 4). We have $\rho_t\kappa_t\rho_t\kappa_t = c\rho_t\kappa_t$.

The right ideal $\rho_t\kappa_t \mathbb{Q}\mathfrak{S}_n$ of $\mathbb{Q}\mathfrak{S}_n$ (which is isomorphic to S^μ) has a complementary right ideal U, by Maschke's Theorem.

Multiplication on the left by $\rho_t\kappa_t$ gives a linear transformation of $\mathbb{Q}\mathfrak{S}_n$. Taking a basis for $\rho_t\kappa_t \mathbb{Q}\mathfrak{S}_n$, followed by a basis of U, this linear transformation is represented by the matrix

$$\dim S^\mu \left\{ \left[\begin{array}{ccc|c} c & & & \\ & \ddots & & 0 \\ & & c & \\ \hline & * & & 0 \end{array}\right] \right.$$

On the other hand, taking the natural basis $\{\pi \mid \pi \in \mathfrak{S}_n\}$ for $\mathbb{Q}\mathfrak{S}_n$, the linear transformation is represented by a matrix with 1's down the diagonal, since the identity permutation occurs with coefficient 1 in the product $\rho_t\kappa_t$.

A comparison of traces gives $c \dim S^\mu = n!$ By the Hook Formula for the dimension of S^μ, $c = \Pi(\text{hook lengths in } [\mu])$.

Since $\{t\}\kappa_t\rho_t \pi = \{t\pi\}\kappa_{t\pi} \rho_{t\pi}$, the first part of the Lemma and Corollary 8.10 show that we may give:

23.3 DEFINITION Suppose that F is the field of p elements. Let θ be the non-zero element of $\operatorname{Hom}_{F\mathfrak{S}_n}(M^\mu, S^\mu)$ given by

$$\theta : \{t\} \to \left(\frac{1}{g^\mu}, \{t\}\kappa_t\rho_t\right)_p$$

where this means that the image of $\{t\}$ is obtained from the vector $\frac{1}{g^\mu}, \{t\}\kappa_t\rho_t$ in $S^\mu_{\mathbb{Q}}$ by reducing all the tabloid coefficients modulo p.

23.4 THEOREM

(i) <u>If $\operatorname{Im}\theta \subset S^\mu$, equivalently if $\operatorname{Ker}\theta \supset S^{\mu\perp}$, then S^μ is reducible</u>

(ii) <u>If $\operatorname{Im}\theta = S^\mu$, equivalently if $\operatorname{Ker}\theta = S^{\mu\perp}$, and if $\operatorname{Hom}_{F\mathfrak{S}_n}(S^\mu, S^\mu) \cong F$, then S^μ is irreducible.</u>

<u>Proof</u>: If $F = \mathbb{Q}$, the the homomorphism ϕ defined by

$$\{t\}\phi = \frac{1}{g^{\mu'}}\{t\}\kappa_t\rho_t$$

sends $\{t\}\kappa_t$ to a non-zero multiple of itself, by Lemma 23.2(ii). Therefore dim Ker ϕ = dim $S_{\mathbb{Q}}^{\mu\perp}$, and by the Submodule Theorem, Ker $\phi = S_{\mathbb{Q}}^{\mu\perp}$. By Lemma 8.14, Ker $\Theta \supseteq S^{\mu\perp}$, when we work over the field of p elements. Therefore, Ker $\Theta \supset S^{\mu\perp}$ if and only if Im $\Theta \subset S^{\mu}$.

The first part of the Theorem is now trivial, since Im Θ is a proper submodule of S^{μ} in this case.

If Ker $\Theta = S^{\mu\perp}$, then Θ gives an isomorphism between $M^{\mu}/S^{\mu\perp}$ and S^{μ}, and result (ii) follows from Lemma 23.1.

23.5 THEOREM <u>Suppose that μ is p-regular. Then S^{μ} is reducible if and only if p divides the integer</u>

$$\{\Pi \text{ (hook lengths in } [\mu])\}/g^{\mu'}$$

<u>Proof</u>: The last Theorem and Corollary 13.17 show that S^{μ} is reducible if and only if Ker $\Theta \supset S^{\mu\perp}$. But, since μ is p-regular, $M^{\mu}/S^{\mu\perp}$ has a unique minimal submodule $(S^{\mu} + S^{\mu\perp})/S^{\mu\perp}$ (by Theorem 4.9). Therefore, S^{μ} is reducible if and only if Ker $\Theta \supset S^{\mu}$.

$$\text{But } \{t\}\kappa_t\,\Theta = \left(\frac{1}{g^{\mu'}}\{t\}\kappa_t\rho_t\kappa_t\right)_p = \left(\frac{\Pi(\text{hook lengths in } [\mu])}{g^{\mu'}}\{t\}\kappa_t\right)_p$$

by Lemma 23.2 (ii). Since S^{μ} is a cyclic module, S^{μ} is reducible if and only if p divides the integer $\frac{\Pi(\text{hook lengths in } [\mu])}{g^{\mu'}}$.

23.6 EXAMPLES (i) If p does not divide Π(hook lengths in $[\mu]$), then (μ is p-regular and) S^{μ} is irreducible. This is just the case where μ is in a block of defect 0 (cf. The Hook Formula).

(ii) If both μ and μ' are p-regular, then from Corollary 10.5, p does not divide $g^{\mu'}$. Thus S^{μ} is reducible if and only if p divides Π(hook lengths in $[\mu]$). For instance, S^{μ} is reducible of $\mu = ((p-1)^x)$ where $1 < x < p$.

(iii) If $\mu = (3,2)$ and $t = \begin{matrix}1&2&3\\4&5\end{matrix}$, then direct computation shows

that $\{t\}\kappa_t\rho_t$ =

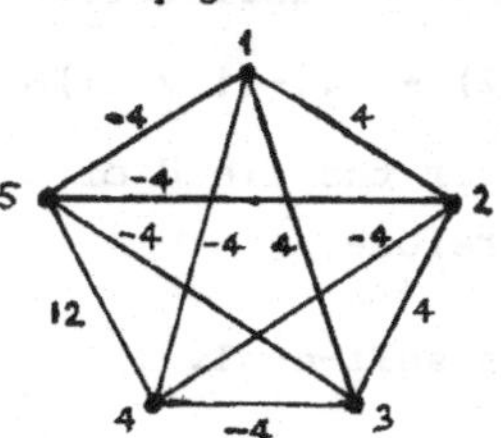

The g.c.d. of the edge coefficients is 4, so $g^{\mu'} = 4$. But the product of the hook lengths in $[\mu]$ is 24, so S^μ is reducible if and only if char F = 2 or 3. When char F = 2, $\{t\}\theta$ is the vector called Γ in Example 5.2, and when char F = 3 , $\{t\}\theta = -\Gamma(-4) - \Gamma(-5)$.

23.7 THEOREM <u>Suppose that μ is a hook partition, and let S^μ be defined over the field of p elements. Then S^μ is irreducible if and only if one of the following holds:</u>

(i) <u>$\mu = (n)$ or (1^n)</u>

(ii) <u>$p \nmid n$ and $\mu = (n-1,1)$ or $(2,1^{n-2})$</u>

(iii) <u>$p \nmid n$ and $p \neq 2$.</u>

<u>Proof</u>: Since $S^{(n)}$ and $S^{(1^n)}$ have dimension 1, they are certainly irreducible. Thus, we may assume that $\mu = (x,1^y)$ with $x > 1$, $y > 0$ and $x + y = n$.

$$\text{Let } t = \begin{matrix} 1 & (y+2)\ldots(y+x) \\ 2 & \\ \cdot & \\ \cdot & \\ (y+1) & \end{matrix}$$

and let $\bar{\kappa}_t = \Sigma\{(\text{sgn }\sigma)\sigma \mid \sigma \in \mathfrak{S}_{\{2,3,\ldots,y+1\}}\}$. Then

$$\kappa_t = (1 - (12) - (13) - \ldots - (1,y+1))\bar{\kappa}_t .$$

For the moment, work over $\mathbb{Q}$. Then

$$\{t\}\kappa_t\rho_t\bar{\kappa}_t = \{t\}\kappa_t\bar{\kappa}_t\rho_t = y!\{t\}\kappa_t\rho_t$$

Therefore,

$$y!\{t\}\kappa_t\rho_t(1 - (12) - \ldots - (1,y+1)) = \{t\}\kappa_t\rho_t\kappa_t$$

$$= \Pi(\text{hook lengths in } [\mu])\{t\}\kappa_t, \text{ by 23.2}$$

$$= (x-1)!\ y!(x+y)\{t\}\kappa_t .$$

But $g^{\mu'} = (x-1)!$ by Lemma 10.4, and so

$$\frac{1}{g^{\mu'}}\{t\}\kappa_t\rho_t(1 - (12) - \ldots - (1,(y+1)) = (x+y)\{t\}\kappa_t .$$

Let θ be the homomorphism of definition 23.3. Then

$$\{t\}(1 - (12) - \ldots - (1,y+1))\theta = (x+y)\{t\}\kappa_t,$$

where we are now working over the field of p elements. This shows that if $p \nmid n$, Im $\theta = S^\mu$. Therefore,

<u>23.8</u> If $p \nmid n$, $S^{(x,1^y)}$ is self-dual.

But $\mathrm{Hom}_{F\mathfrak{S}_n}(S^\mu,S^\mu) \cong F$ if $p \neq 2$ or if $\mu = (n-1,1)$, by Corollary 13.17. Using Lemma 23.1, S^μ is irreducible in the cases where $p \nmid n$ and $p \neq 2$ or $\mu = (n-1,1)$ (also when $\mu = (2,1^{n-2})$, by Theorem 8.15).

Next suppose that $p \mid n$. Then

$$\{t\}(1 - (12) - \ldots - (1,y+1)) \in \mathrm{Ker}\,\Theta$$

$$\text{Let } t^* = \begin{matrix} (y+x) & (y+x-1) \ldots (y+2) & 1 \\ 2 & & \\ \cdot & & \\ \cdot & & \\ (y+1) & & \end{matrix}$$

Since $x > 1$, all the tabloids in e_{t^*} have 1 in the first row. Hence $\{t\} = \{t^*\}$ is the unique tabloid involved in both e_{t^*} and $\{t\}(1 - (12) - \ldots - (1,y+1))$, and so

$$< \{t\}(1 - (12) - \ldots - (1,y+1)), e_{t^*} > = 1.$$

Therefore, $\{t\}(1 - (12) - \ldots - (1,y+1)) \in \mathrm{Ker}\,\Theta \setminus S^{\mu\perp}$, and Theorem 23.4 proves S^μ is reducible in this case, where $p \mid n$.

Finally, we prove that S^μ is reducible when $\mu = (x,1^y)$ with $x > 1$, $y > 1$ and $p = 2$. By Theorem 8.15, we may assume that $x \geq y$. Observe that

$$[x][y] = [x+y] + [x+y-1,1] + \ldots + [x,y]$$

$$\text{and} \quad [x][1^y] = [x+1,1^{y-1}] + [x,1^y]$$

by the Littlewood-Richardson Rule. But when $p = 2$, $\chi^{(y)}$ and $\chi^{(1^y)}$ are the same 2-modular character, and thus

$$\chi^{(x+1,1^{y-1})} + \chi^{(x,1^y)} = \chi^{(x+y)} + \chi^{(x+y-1,1)} + \ldots + \chi^{(x,y)}$$

as a 2-modular character. Whence, by induction,

$$\chi^{(x,1^y)} = \chi^{(x,y)} + \chi^{(x+2,y-2)} + \chi^{(x+4,y-4)} + \ldots$$

and so $\chi^{(x,1^y)}$ is certainly a reducible 2- modular character.

<u>Remark</u>: The last part of the proof shows that

$(n),(n-2,2),(n-4,4),\ldots$ are in the same 2-block,

and $(n-1,1),(n-3,3),(n-5,5),\ldots$ are in the same 2-block of $\mathfrak{S}_n$ (see Theorem 21.9). When n is even, all the 2-part partitions of n are in the same 2-block of $\mathfrak{S}_n$, since Example 5.1 proves that (n) and (n-1,1) are in the same 2-block (see also, Example 21.10). When n is odd, the 2-part partitions of n lie in two different 2-blocks, since Lemma 21.12 shows that (n) and (n-1,1) are in different 2-blocks.

Theorem 23.7 will help us in our first result in the next chapter on the decomposition matrices of $\mathfrak{S}_n$. For hook partitions, $g^{\mu'}$ is easy to calculate; unfortunately, this is not the case for other types of partition, for example:

23.9 LEMMA <u>If $\mu = (x,y)$, then</u>

$$g^{\mu'} = y!\ \text{g.c.d.}\ \{x!,\ (x-1)!1!,\ (x-2)!2!,\ldots,(x-y)!y!\}$$

<u>Proof</u>: Let t_1 and t_2 be μ'-tableaux. Let

X_{ij} = {k|k belongs to the ith column of t_1 and to the jth column of t_2}

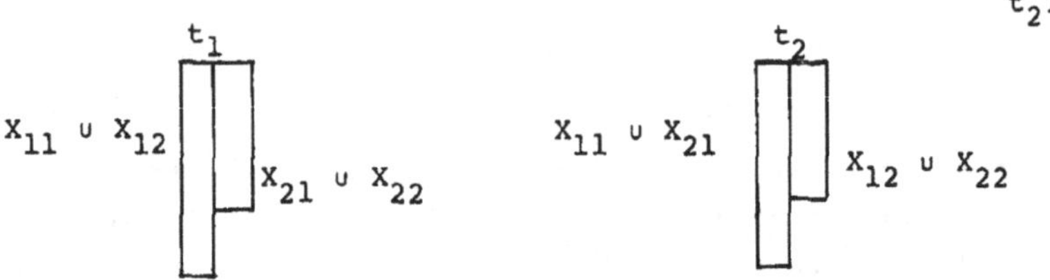

The polytabloids e_{t_1} and e_{t_2} in $S^{\mu'}_{\mathbb{Q}}$ have the tabloid $\{t_3\}$ in common if an only if no two numbers from any one of the sets $X_{11} \cup X_{12}$, $X_{21} \cup X_{22}$, $X_{11} \cup X_{21}$, $X_{12} \cup X_{22}$ are in the same row of $\{t_3\}$. Any row of $\{t_3\}$ must contain a number from X_{12} and a number from X_{21} or no numbers from $X_{12} \cup X_{21}$. Therefore, $< e_{t_1}, e_{t_2} > = 0$ unless $|X_{12}| = |X_{21}|$.

Suppose now that $|X_{12}| = |X_{21}|$. The tabloid $\{t_3\}$ is common to e_{t_1} and e_{t_2} if and only if each of the first y rows of $\{t_3\}$ is occupied by just one number from $X_{21} \cup X_{22}$ and each row containing a number from X_{21} contains a number from X_{12}. Thus, e_{t_1} and e_{t_2} have $y!\ |X_{12}|!\ (x - |X_{12}|)!$ common tabloids.

Assume that the tabloid representative t_3 for the common tabloid $\{t_3\}$ has been chosen such that $t_3 = t_1\pi_1$ for some π_1 in the column stabilizer of t_1. Let σ be the permutation in the row stabilizer of t_3 interchanging each number in X_{12} with a number in X_{21}, leaving the other numbers fixed. Then $t_3\sigma = t_2\pi_2$ for some π_2 in the column stabilizer of t_2, and $\text{sgn}\ \sigma = (-1)^{|X_{12}|}$. Therefore, $t_1\ \pi_1\sigma\ \pi_2^{-1} = t_2$, and $(\text{sgn}\ \pi_1)$ $(\text{sgn}\ \pi_2)$ depends only on t_1 and t_2 and not on $\{t_3\}$. But $\{t_3\} = \{t_1\}\pi_1$ $= \{t_2\}\pi_2$, and hence

$$< e_{t_1}, e_{t_2} > = \pm y!\ |X_{12}|!\ (x - |X_{12}|)!$$

By definition, $g^{\mu'}$ is the greatest common divisor of such integers, and, since $0 \le |X_{12}| \le y$, the Lemma is proved.

23.10 EXAMPLES

(i) If $\mu = (5,2)$, then $g^{\mu'} = 2!\ \text{g.c.d.}(5!,4!1!,3!2!) = 2^3.3$. But Π(hook lengths in $[\mu]$) $= 2^3,3^2,5$. Therefore, $S^{(5,2)}$ is reducible if and only if the ground field has characteristic 3 or 5.

(ii) Similarly, $S^{(5,3)}$ is reducible if and only if the ground field has characteristic 2 or 5.

(iii) If $p \neq 7$, $S^{(5,1^2)}$ is self-dual, by 23.8. Now let the ground field have characteristic $p = 2$. Then the first example proves $S^{(5,2)}$ is irreducible, and Example 21.8(ii) shows that $S^{(5,1^2)}$ has composition factors isomorphic to $S^{(5,2)}$ and $S^{(7)}$. Since $S^{(5,1^2)}$ is self-dual, these factors can occur in either order, and so $S^{(5,1^2)}$ is decomposable over a field of characteristic 2.

The last Example proves that the hypotheses cannot be omitted in 13.17, 13.18, 23.1 or 23.4.

23.11 DEFINITION The <u>p-power diagram</u> $[\mu]^p$ for μ is obtained by replacing each integer h_{ij} in the hook graph for μ by $\nu_p(h_{ij})$.

23.12 EXAMPLE If $\mu = (8,5,2)$, then the hook graph is

$$\begin{matrix} 10 & 9 & 7 & 6 & 5 & 3 & 2 & 1 \\ 6 & 5 & 3 & 2 & 1 & & & \\ 2 & 1 & & & & & & \end{matrix}$$

$$\text{and } [\mu]^3 = \begin{matrix} 0 & 2 & 0 & 1 & 0 & 1 & 0 & 0 \\ 1 & 0 & 1 & 0 & 0 & & & \\ 0 & 0 & & & & & & \end{matrix}$$

$$\text{and } [\mu]^2 = \begin{matrix} 1 & 0 & 0 & 1 & 0 & 0 & 1 & 0 \\ 1 & 0 & 0 & 1 & 0 & & & \\ 1 & 0 & & & & & & \end{matrix}$$

We now classify the irreducible Specht modules corresponding to 2-part partitions.

23.13 THEOREM <u>Suppose $\mu = (x,y)$ is p-regular (i.e. if $p = 2$, we assume $x \neq y$). Then S^μ defined over the field of p elements, is reducible if and only if some column of $[\mu]^p$ contains two different numbers.</u>

<u>Proof</u>: The hook lengths h_{ij} for $[\mu]$ are given by

$$\begin{aligned} h_{1j} &= x - j + 2 && \text{for } 1 \le j \le y \\ h_{1j} &= x - j + 1 && \text{for } y < j \le x \\ h_{2j} &= y - j + 1 && \text{for } 1 \le j \le y. \end{aligned}$$

If there is a j with $\nu_p(h_{1j}) \neq \nu_p(h_{2j})$, consider the largest j with this property and let $\nu_p(h_{2j}) = r$. Then $j + p^r \le y + 1$ and

$$\nu_p(h_{1i}) = \nu_p(h_{2i}) < r \quad \text{for } j + 1 \le i < j + p^r$$

But $\{h_{1i} \mid j \le i < j + p^r\}$ is a set of p^r consecutive integers, so $\nu_p(h_{1j}) \ge r = \nu_p(h_{2j})$. Since $\nu_p(h_{1j}) \ne \nu_p(h_{2j})$, we have $\nu_p(x - j+2) > \nu_p(y - j+1)$. Writing $b = x - j+2$ and noting that $\nu_p(b) > \nu_p(b - x+y - 1)$ if and only if $\nu_p(b) > \nu_p(x - y+1)$, this proves

<u>23.14</u> Some column of $[x,y]^p$ contains two different numbers if and only if there is an integer b with $x - y+2 \le b \le x+1$ and $\nu_p(b) > \nu_p(x - y+1)$

Now, $\Pi(\text{hook lengths in } [x,y]) = (y!(x+1)!)/(x - y+1)$ and $g^{\mu'} = y!\ \text{g.c.d.}\{x!, (x - 1)!1!, \ldots, (x - y)!y!\}$ by Lemma 23.9, so Theorem 23.5 proves that S^μ is reducible if and only if p divides

$$\frac{x+1}{x - y+1}\ \text{l.c.m.}\ \{\binom{x}{x}, \binom{x}{x-1}, \ldots, \binom{x}{x-y}\}\ .$$

Since $(x+1)\binom{x}{b-1} = b\binom{x+1}{b}$,

<u>23.15</u> $S^{(x,y)}$ is reducible if and only if there is an integer b with $x - y+1 \le b \le x+1$ and $\nu_p\ \{\frac{b}{x - y+1}\binom{x+1}{b}\} > 0$.

Comparing 23.14 and 23.15, we see that $S^{(x,y)}$ is reducible if some column of $[x,y]^p$ contains two different numbers.

On the other hand, suppose that no column of $[x,y]^p$ contains different numbers. Then, for every b with $x - y+2 \le b \le x+1$, $\nu_p(b) \le \nu_p(x - y+1)$.

Let
$$x - y+1 = a_r p^r + a_{r+1}p^{r+1} + \ldots + a_s p^s \qquad (0 \le a_i < p,\ a_r \ne 0 \ne a_s).$$

Then
$$x - y+1 < (a_{r+1} + 1)p^{r+1} + a_{r+2}p^{r+2} + \ldots + a_s p^s$$

and $\nu_p((a_{r+1} + 1)p^{r+1} + \ldots + a_s p^s) > \nu_p(x - y+1)$. Thus our supposition gives $x+1 < (a_{r+1} + 1)p^{r+1} + \ldots + a_s p^s$. Therefore

$$x+1 = c_o + c_1 p + \ldots + c_r p^r + a_{r+1}p^{r+1} + \ldots + a_s p^s \qquad (0 \le c_i < p)$$

and if $x - y + 1 \le b \le x+1$, then

$$b = b_q p^q + b_{q+1}p^{q+1} + \ldots + b_r p^r + a_{r+1}p^{r+1} + \ldots + a_s p^s \qquad (0 \le b_i < p,\ b_q \ne 0)$$

Therefore,
$$x+1 - b = c_o + c_1 p + \ldots + c_{q-1}p^{q-1} + d_q p^q + \ldots + d_r p^r \qquad (0 \le d_i < p)$$

where $d_q p^q + \ldots + d_r p^r = c_q p^q + \ldots + c_r p^r - b_q p^q - \ldots - b_r p^r$

By Lemma 22.2,

$$\begin{aligned}
\nu_p\binom{x+1}{b} &= \{\sigma_p(b) + \sigma_p(x+1-b) - \sigma_p(x+1)\}/(p-1) \\
&= (b_q + \ldots + b_r + d_q + \ldots + d_r - c_q - \ldots - c_r)/(p-1) \\
&= \nu_p \binom{c_q p^q + \ldots + c_r p^r}{b_q p^q + \ldots + b_r p^r} \\
&\le r - q, \text{ by Lemma 22.3 (since } b_q \neq 0) \\
&= \nu_p(x - y+1) - \nu_p(b).
\end{aligned}$$

Therefore, for $x - y+1 \le b \le x+1$, $\nu_p\{\frac{b}{x-y+1}\binom{x+1}{b}\} \le 0$ and $S^{(x,y)}$ is irreducible,as required.

23.16 EXAMPLE $S^{(2p-1,p)}$ is irreducible over the field of p elements if and only if $p \neq 2$ (cf. Example 23.10). This is interesting because an earlier author believed, apparently on the evidence of the case $p = 2$, that $S^{(2p-1,p)}$ always has two composition factors, one being the trivial module $D^{(3p-1)}$. Since dim $S^{(2p-1,p)} \equiv 1 \bmod p^3$ for p odd - this follows from the Hook Formula - the mistake would have provided counterexamples to a conjecture of Brauer which states that $\nu_p(|G|/\dim D) \ge 0$ for each p-modular irreducible representation D of a group G.

R.W. Carter has put forward

23.17 CONJECTURE <u>No column of $[\mu]^p$ contains two different numbers if and only if μ is p-regular and S^μ is irreducible over the field of p elements.</u>

It is trivial that $[\mu]^p$ has a column containing two different numbers if μ is p-singular. The author [11] has proved that the given condition is necessary for a p-regular Specht module to be irreducible, and has proved it is sufficient in the case where $p = 2$.

Over the field of 2 elements, it turns out that $S^{(x,x)}$ is irreducible if and only if $x = 1$ or 2 (This is the only 2 part partition not considered in Theorem 23.13). We conjecture that (2,2) is the unique partition μ such that S^μ is irreducible over the field of 2 elements but neither μ nor μ' is 2-regular.

There is no known way of determining the composition factors of the general Specht module when the ground field F has characteristic a prime p. Thus we cannot decide the entries in the decomposition matrix of $\mathfrak{S}_n$, which records the multiplicity of each p modular irreducible representation D^λ (λ p-regular) as a composition factor of S^μ, except in some special cases. The theorems we expound give only <u>partial</u> results.

24.1 THEOREM (Peel [18]) <u>Suppose p is odd.</u>

(i) <u>If $p \nmid n$, all the hook representations of $\mathfrak{S}_n$ remain irreducibl</u> <u>modulo p, and no two are isomorphic</u>.

(ii) <u>If $p \mid n$, part of the decomposition matrix of $\mathfrak{S}_n$ is</u>

$$\begin{array}{lcccccc}
(n) & 1 & & & & & \\
(n-1,1) & 1 & 1 & & & 0 & \\
(n-1,1^2) & & 1 & 1 & & & \\
\vdots & & & & \ddots & & \\
(2,1^{n-2}) & & 0 & & & 1 & 1 \\
(1^n) & & & & & & 1
\end{array}$$

<u>Proof</u>: The result is true for n = 0, so we may assume that it is true for n - 1. Note that

$$\chi^{(x,1^y)} \downarrow \mathfrak{S}_{n-1} = \chi^{(x-1,1^y)} + \chi^{(x,1^{y-1})} \quad \text{if} \quad x > 1,\ y > 0,\ x+y = n.$$

Case (i) p does not divide n.

In view of Theorem 23.7, we need prove only that no two hook representations are isomorphic. But this follows at once, since they have non-isomorphic restrictions to $\mathfrak{S}_{n-1}$.

Case (ii) p divides n.

Suppose $x > 1$, $y > 0$. Then by restricting to $\mathfrak{S}_{n-1}$, $\chi^{(x,1^y)}$ has at most two modular constituents, and therefore precisely two, by Theorem 23.7. Let ϕ_x^+ be the modular constituent of $\chi^{(x,1^y)}$ satisfying $\phi_x^+ \downarrow \mathfrak{S}_{n-1} = \chi^{(x-1,1^y)}$ and ϕ_x^- be that satisfying $\phi_x^- \downarrow \mathfrak{S}_{n-1} = \chi^{(x,1^{y-1})}$

(and let $\phi_n^- = 0$ and $\phi_1^+ = 0$). We must show that for every x, $\phi_{x-1}^- = \phi_x^+$. no other equalities can hold because there are different restrictions to $\mathfrak{S}_{n-1}$.

The following relation between characters holds on all classes except (n), in particular on all p-regular classes:

$$\chi^{(n)} - \chi^{(n-1,1)} + \chi^{(n-2,1^2)} - \ldots \pm \chi^{(1^n)} = 0.$$

(This comes from Theorem 21.7 or direct from Theorem 21.4, by using the ordinary character orthogonality relations).

In terms of modular characters; we have

$$\phi_n^+ - (\phi_{n-1}^- + \phi_{n-1}^+) + (\phi_{n-2}^- + \phi_{n-2}^+) - \ldots \pm \phi_1^- = 0.$$

If some ϕ_{x-1}^- were not equal to ϕ_x^+, then ϕ_{x-1}^- would appear just once in this relation, contradicting the fact that the modular irreducible characters of a group are linearly independent.

From now on, we shall label the rows of our decomposition matrices by partitions, and the columns by p-regular partitions. Thus the entry in the μth row and λth column is the multiplicity of D^λ as a composition factor of S^μ over a field of characteristic p. Omitted entries in decomposition matrices are zero. We write χ^μ for the p-modular character of S^μ and ϕ^λ for the p-modular character of D^λ.

24.2 EXAMPLE When p = 3, the decomposition matrix of $\mathfrak{S}_5$ is

	(5)	(4,1)	(3,2)	(3,1²)	(2²,1)
(5)	1				
(4,1)		1			
(3,2)		1	1		
(3,1²)				1	
(2²,1)	1				1
(2,1³)					1
(1⁵)			1		

<u>Proof</u>: The rows corresponding to (5), (4,1) and (3,1²) come from Theorem 24.1.

Taking $[\nu] = [2]$ and $r = 3$ in Theorem 21.7, we find that

$$\chi^{(5)} - \chi^{(2^2,1)} + \chi^{(2,1^3)} = 0 \text{ on 3-regular classes.}$$

But $\chi^{(5)}$ and $\chi^{(2,1^3)}$ are irreducible and inequivalent, by Theorem 24.1. Thus, $\chi^{(2^2,1)}$ has precisely two factors. Since one of these must be $\phi^{(2^2,1)}$, it follows that

$$\chi^{(2^2,1)} = \phi^{(5)} + \phi^{(2^2,1)}$$

$$\text{and} \quad \chi^{(2,1^3)} = \phi^{(2^2,1)}.$$

The rest of the matrix is similarly deduced from the equation:

$$\chi^{(1^5)} - \chi^{(3,2)} + \chi^{(4,1)} = 0 \text{ on 3-regular classes.}$$

24.3 EXAMPLE When p = 3, the decomposition matrix of $\mathfrak{S}_6$ is that given in the Appendix.

<u>Proof</u>: First note that $\chi^{(4,2)}$ and $\chi^{(2^2,1^2)}$ are irreducible by Example 23.6(i).

By Theorem 24.1, part of the matrix is

	(6)	(5,1)	(4,1²)		
(6)	1				
(5,1)	1	1			
(4,1²)		1	1		
(3,1³)			1	1	
(2,1⁴)				1	1
(1⁶)					1

Applying Theorem 21.7, with r = 3 and $[\nu] = [3]\ [2,1]$ and $[1^3]$ in turn we get,

$$\chi^{(6)} + \chi^{(3^2)} - \chi^{(3,2,1)} + \chi^{(3,1^3)} = 0$$
$$\chi^{(5,1)} - \chi^{(3^2)} - \chi^{(2^3)} + \chi^{(2,1^4)} = 0$$
$$\chi^{(4,1^2)} - \chi^{(3,2,1)} + \chi^{(2^3)} + \chi^{(1^6)} = 0$$

on 3-regular classes. These equations, together with

$$\chi^{(6)} - \chi^{(5,1)} + \chi^{(4,1^2)} - \chi^{(3,1^3)} - \chi^{(2,1^4)} - \chi^{(1^6)} = 0$$

enable us to deduce that the remaining two columns above should be labelled (3,2,1) and (3^2), respectively, and the equations let us write $\chi^{(3^2)}$, $\chi^{(3,2,1)}$ and $\chi^{(2^3)}$ in terms of $\phi^{(6)}$, $\phi^{(5,1)}$, ..., in the way shown in the complete decomposition matrix in the Appendix.

Note that Examples 24.2 and 24.3 have been computed without using the Nakayama Conjecture, and without resorting to induction (except where it is implicit in Theorem 24.1). We agree that it is quicker to deduce the decomposition matrix of $\mathfrak{S}_6$ from that of $\mathfrak{S}_5$ using the Branching Theorem and block theory, but this traditional method of finding decomposition matrices fails to determine the factors of $S^{(2p-1,p)}$, even for p = 2 (cf. Example 23.16), and very rapidly leads to further ambiguities.

It seems to us that if a method is eventually devised for finding the decomposition matrices of $\mathfrak{S}_n$, it will include information concernin the order of the factors of each Specht module, as well as the multiplicities of the composition factors. For this line of attack, the most useful Theorems we know are Theorem 13.13, giving a basis of $\text{Hom}_{F\mathfrak{S}_n}(S^\lambda, M^\mu)$ and Corollary 17.18, describing S^μ as a kernel intersection. It is unfortunate that these two results look rather ugly, and that the notation which has to be used obscures the simplicity of their application, but we embark upon the task of employing them.

We return to the notation of section 13, where M^μ is described as th space spanned by λ-tableaux of type μ. The remarks following 17.8 and 17.10 show that the homomorphism $\psi_{i,v}$ acts on M^μ by sending a tableau T to the sum of all the tableaux obtained by changing all but v (i+1)'s to i's.

e.g. $\psi_{1,1}$: $\begin{matrix}1&1&1&2&2\\2&3&3\end{matrix} \to \begin{matrix}1&1&1&1&1\\2&3&3\end{matrix} + \begin{matrix}1&1&1&1&2\\1&3&3\end{matrix} + \begin{matrix}1&1&1&2&1\\1&3&3\end{matrix}$

The first result we prove could be subsumed in Theorem 24.6, but we present the special case to help the reader become familiar with the relevant ideas.

24.4 THEOREM <u>Over a field of prime characteristic p, S^μ has a submodule isomorphic to the trivial $\mathfrak{S}_n$-module $S^{(n)}$ if and only if for all i, $\mu_i \equiv -1 \bmod p^{z_i}$ where $z_i = \ell_p(\mu_{i+1})$.</u>

<u>Proof:</u> By Theorem 13.13 (or trivially) there is, to within a scalar multiple, a unique element Θ_T in $\mathrm{Hom}_{F\mathfrak{S}_n}(S^{(n)}, M^\mu)$. T is the semistandard (n)-tableau of type μ, and Θ_T sends {t} to the sum of the (n)-tableaux of type μ .
e.g. if $\mu = (3,2)$, then
$\{t\}\Theta_T$ = 1 1 1 2 2 + 1 1 2 1 2 + 1 1 2 2 1 + 1 2 1 1 2 + 1 2 1 2 1 + 1 2 2 1 1 + 2 1 1 1 2 + 2 1 1 2 1 + 2 1 2 1 1 + 2 2 1 1 1 .

Now, the crucial step is that when T_1 is an (n)-tableau of type $(\mu_1, \mu_2, \ldots, \mu_{i-1}, \mu_i + \mu_{i+1} - v, v, \mu_{i+2}, \ldots)$ there are

$$\binom{\mu_i + \mu_{i+1} - v}{\mu_{i+1} - v}$$

tableaux row equivalent to T in which all but v (i+1)'s can be changed to i's to give T_1
e.g. 1 1 1 1 1 comes from $\binom{5}{2}$ tableaux above, by changing all the 2's to 1's, and each of 1 1 1 1 2, 1 1 1 2 1, 1 1 2 1 1, 1 2 1 1 1 , 2 1 1 1 1 comes from $\binom{4}{1}$ tableaux by changing all except one 2 to 1.

Therefore, $\{t\}\Theta_T$ belongs to $\bigcap_{v=0}^{\mu_{i+1}-1} \ker \psi_{i,v}$ if and only if each of

$$\binom{\mu_i + \mu_{i+1}}{\mu_{i+1}}, \binom{\mu_i + \mu_{i+1} - 1}{\mu_{i+1} - 1}, \ldots, \binom{\mu_i + 1}{1}$$

is divisible by p. This is equivalent to $\mu_i \equiv -1 \bmod p^{z_i}$ where $z_i = \ell_p(\mu_{i+1})$, by Corollary 22.5. Thus, Corollary 17.18 shows that $\{t\}\Theta_T$ belongs to S^μ if and only if this congruence holds for all $i \geq 1$.

24.5 EXAMPLES (i) $S^{(8,2,2,1)}$ contains a trivial submodule if and only if the ground field F has characteristic 3.

(ii) $S^{(5,2)}$ does not contain a trivial submodule if char F = 2.

(iii) $S^{(p-1,p-1,\ldots,p-1,r)}$ contains a trivial submodule if char F = p, and r < p. Write n = x(p - 1)+r. Then $((x+1)^r, x^{p-1-r})$ is the partition μ' conjugate to $\mu = ((p-1)^x, r)$. Since $\mathrm{Hom}_{F\mathfrak{S}_n}(S^{(n)}, S^{\mu}) \neq 0$, and $S^{\mu} \otimes S^{(1^n)}$ is isomorphic to the dual of $S^{\mu'}$ it follows that $\mathrm{Hom}_{F\mathfrak{S}_n}(S^{\mu'}, S^{(1^n)}) \neq 0$. By construction, $S^{\mu'}$ is p-regular, so μ' is the unique partition of n such that $D^{\mu'} \cong S^{(1^n)}$ (Remember that $D^{\mu'}$ is the unique top composition factor of $S^{\mu'}$). Compare Example 24.2, where $S^{(1^5)} \cong D^{(3,2)}$.

(iv) Consulting the decomposition matrices in the Appendix, we see that $S^{(4,2)}$ has a trivial composition factor for p = 2, but $S^{(4,2)}$ does not have a trivial <u>bottom</u> composition factor, by Theorem 24.4.

It is interesting to see that for any given λ and μ, we can use Theorem 13.13 and Corollary 17.18 to determine whether or not $\mathrm{Hom}_{F\mathfrak{S}_n}(S^{\lambda}, S^{\mu})$ is zero (except in the rather uninteresting case where char F = 2 and λ is 2-singular), for we may list the semistandard homomorphisms from M^{λ} into M^{μ} and then test whether some linear combination of them sends $\{t\}\kappa_t$ into the kernel intersection of Corollary 17.18. This is a tedious task, but not altogether impossible, even for fairly large partitions. For example, after a little practice on small partitions, the reader should have no difficulty using the technique of Theorem 24.6 below to prove that $\mathrm{Hom}_{F\mathfrak{S}_n}(S^{\lambda}, S^{(10,5,3,)}) = 0$ when char F = 3 and λ = (16,2), (13,5) or (10,8). Using the Nakayama Conjecture, this proves that $S^{(10,5,3)}$ is irreducible over fields of characteristic 3 (cf. Carter's Conjecture 23.17).

When applying Theorem 13.13 and Corollary 17.18, we are usually interested in the case where S^{λ} is p-regular, since then $\mathrm{Hom}_{F\mathfrak{S}_n}(S^{\lambda}, S^{\mu}) \neq 0$ implies that D^{λ} is a composition factor of S^{μ}. Unfortunately, a completeclassification of the cases where $\mathrm{Hom}_{F\mathfrak{S}_n}(S^{\lambda}, S^{\mu})$ is non-zero is not sufficient to determine the decomposition matrix of $\mathfrak{S}_n$; in Example 24.5(iv) $D^{(6)}$ is a factor of $S^{(4,2)}$ over the field F of 2 elements, but $\mathrm{Hom}_{F\mathfrak{S}_n}(S^{(6)}, S^{(4,2)}) = 0$. Even so, sometimes a modification of the method is good enough to classify all the composition factors of S^{μ}; see Theorem 24.15 below, for example.

In section 13 we saw that there is much choice in the way we define a semistandard λ-tableau of type μ. It turns out that it is often most useful to consider tableaux where the numbers are non-increasing along the rows and strictly decreasing down the columns; we shall call such a tableau <u>reverse semistandard</u>. The second part of the next Theorem probably classifies all cases where there is a reverse semistandard homomorphism in $\mathrm{Hom}_{F\mathfrak{S}_n}(S^\lambda, S^\mu)$. When considering linear combinations of more than one semistandard homomorphism, the situation becomes horribly complicated!

24.6 THEOREM <u>Assume that λ and μ are (proper) partitions of n and that char $F = p$. Suppose that T is a reverse semistandard λ-tableaux of type μ, and let N_{ij} be the number of i's in the jth row of T.</u>

(i) <u>If for all $i \geq 2$ and $j \geq 1$, $N_{i-1,j} \equiv -1 \bmod p^{a_{ij}}$ where $a_{ij} = \ell_p(N_{ij})$, then θ_T belongs to $\mathrm{Hom}_{F\mathfrak{S}_n}(M^\lambda, S^\mu)$ and $\mathrm{Ker}\,\theta_T \subseteq S^{\lambda\perp}$.</u>

(ii) <u>If for all $i \geq 2$ and $j \geq 1$, $N_{i-1,j} \equiv -1 \bmod p^{b_{ij}}$ where $b_{ij} = \min\{\ell_p(N_{ij}), \ell_p(\sum_{m=1}^{i-1}(\lambda_{j+m-1} - \sum_{s=j}^{\infty} N_{ms}))\}$, then $\hat{\theta}_T$ is a non-zero element of $\mathrm{Hom}_{F\mathfrak{S}_n}(S^\lambda, S^\mu)$.</u>

<u>Proof</u>: Since T is reverse semistandard, $\mathrm{Ker}\,\theta_T \not\supseteq S^\lambda$ by Lemma 13.11 and the Remark following Corollary 13.14. Therefore, $\mathrm{Ker}\,\theta_T \subseteq S^{\lambda\perp}$ by the Submodule Theorem.

Let t be the λ-tableau used to define the $\mathfrak{S}_n$ action on M^μ. Then $\{t\}\theta_T$ is, by definition, the sum of the λ-tableaux of type μ which are row equivalent to T.

Let $i \geq 2$, $0 \leq v \leq \mu_i - 1$. Since $\sum_{j=1}^{\infty} N_{ij} = \mu_i$, we may choose $v_1, v_2, \ldots$ such that $0 \leq v_j \leq N_{ij}$ for each j and $\Sigma\, v_j = v$. Choose a tableau T_1 row equivalent to T, and for each j change all except v_j i's in the jth row of T_1 into (i-1)'s. Let T_2 be the resulting tableau. By definition, each tableau T_2 involved in $\{t\}\theta_T\psi_{i-1,v}$ is constructed in this way, and T_2 appears in $\{t\}\theta_T\,\psi_{i-1,v}$ from

$$\prod_{j=1}^{\infty} \binom{N_{i-1,j} + N_{ij} - v_j}{N_{ij} - v_j}$$

different tableaux row equivalent to T.

Since $\sum_{j=1}^{\infty} N_{ij} = \mu_i > v = \sum_{j=1}^{\infty} v_j$, there is an integer k with

$$0 \leq v_k < N_{ik} .$$

If for all j $N_{i-1,j} \equiv -1 \bmod p^{a_{ij}}$ then

$$\binom{N_{i-1,k} + N_{ik} - v_k}{N_{ik} - v_k}$$

is divisible by p, by Corollary 22.5. Thus if the hypothesis of part (i) of the Theorem holds, Corollary 17.18 proves that $M^\lambda\theta_T \subseteq S^\mu$ as required.

Under the hypothesis of part (ii), it again follows that

$\{t\}\kappa_t\ \psi_{i-1,v}$ does not involve T_2, except if

$$N_{ik} - v_k > \sum_{m=1}^{i-1} (\lambda_{k+m-1} - \sum_{s=k}^{\infty} N_{ms}) .$$

But for $m < i - 1$, T_2 has $\sum_{s=k}^{\infty} N_{ms}$ numbers equal to m in rows $k,k+1,\ldots$ since T_2 has come from a tableau row equivalent to T. Similarly, T_2 has at least $\sum_{s=k}^{\infty} N_{i-1,s} + N_{ik} - v_k$ numbers equal to $i - 1$ in rows $k,k+1,\ldots,$ since $N_{ik} - v_k$ i's have been changed to (i-1)'s in row k. Altogether, therefore, T_2 has at least

$$N_{ik} - v_k + \sum_{m=1}^{i-1} \sum_{s=k}^{\infty} N_{ms}$$

numbers less than or equal to i-1 in rows $k,k+1,\ldots$. If we assume that this excedes $\sum_{m=1}^{i-1} \lambda_{k+m-1}$, it follows that some column of T_2 contains two identical numbers. Therefore, T_2 is annihilated by κ_t . This shows that in part (ii) of the Theorem, $\{t\}\Theta_T\psi_{i-1,v}\ \kappa_t = 0$ when $i \geq 2$ and $0 \leq v \leq \mu_i - 1$; thus, $\{t\}\kappa_t\Theta_T$ belongs to S^μ, as we wished to prove

Since $M^\lambda/S^{\lambda\perp}$ is isomorphic to the dual of S^λ, and $S^\lambda \cap S^{\lambda\perp}$ is the unique maximal submodule of S^λ when λ is p-regular we have

24.7 COROLLARY <u>Under the hypothesis of part (i) of Theorem 24.6, every composition factor of S^λ is a composition factor of S^μ. Under the second hypothesis, D^λ is a composition factor of S^μ if λ is p-regular</u>.

There are very many applications of Corollary 24.7. We give just one, but we shall use the Corollary again later to find all the composition factors of Specht modules corresponding to 2-part partitions.

24.8 EXAMPLE (cf. Example 24.3). Let $\mu = (3,2,1)$ and char F = 3. Then all the factors of $S^{(5,1)}$ are factors of S^μ; take T =
```
3 2 2 1 1
1
```

$D^{(3^2)}$ is a factor of S^μ; take T =
```
3 2 2
1 1 1
```

$D^{(4,1^2)}$ is a factor of S^μ; take T =
```
3 2 1 1
2
1
```

Theorem 24.6 also gives

24.9 COROLLARY <u>If for all $i \geq 2$, $\mu_{i-1} - \mu_i \equiv -1 \bmod p^{z_i}$ where $z_i = \ell_p(\mu_i - \mu_{i+1})$, then S^μ is irreducible over a field of characteristic p.</u>

<u>Proof</u>: The unique reverse semistandard μ-tableau T of type μ has $N_{ij} = \mu_{i+j-1} - \mu_{i+j}$. Our hypothesis and the first part of Theorem 24.6 show that Θ_T belongs to $\mathrm{Hom}_{F\mathfrak{S}_n}(M^\mu,S^\mu)$ and $\mathrm{Ker}\ \Theta_T \subseteq S^{\mu\perp}$.

By dimensions, $M^{\mu}/S^{\mu\perp} \cong S^{\mu}$. The parts of μ must be strictly decreasing, so μ is certainly p-regular. The result now follows from Lemma 23.1.

When $p = 2$, it is straightforward to verify that the hypothesis of the above Corollary is equivalent to the statement that no column of the 2-power diagram $[\mu]^2$ contains two different numbers; cf. the comments following the Carter Conjecture 23.17.

To describe another special case of Theorem 24.6, we write $\mu \overset{i}{\to} \lambda$ if we can obtain $[\lambda]$ from $[\mu]$ by moving some number $d \geq 0$ of nodes from the end of the ith row of $[\mu]$ to the end of the $(i-1)$th row of $[\mu]$ and each node is moved through a multiple of $p^{\ell_p(d)}$ spaces. (See Example 24.11).

24.10 COROLLARY <u>Let char $F = p$ and $\mu^{(1)}, \mu^{(2)}, \ldots, \mu^{(r)}$ be (proper) partitions of n with</u>

$$\mu^{(1)} \overset{k}{\to} \mu^{(2)} \overset{k-1}{\to} \mu^{(3)} \overset{k-2}{\to} \cdots \overset{k-r+2}{\to} \mu^{(r)}$$

<u>If $1 \leq a \leq b \leq r$ and $\lambda = \mu^{(b)}$, $\mu = \mu^{(a)}$ then $\mathrm{Hom}_{F\mathfrak{S}_n}(S^{\lambda}, S^{\mu}) \neq 0$.</u>

<u>Proof</u>: We may suppose that $a = 1$ and $b = r$, since otherwise we may restrict our attention to the sequence $\mu^{(a)} \to \cdots \to \mu^{(b)}$.

Let d_j be the number of nodes moved in $\mu^{(k-j+1)} \overset{j}{\to} \mu^{(k-j+2)}$ (defining $d_j = 0$ if $j > k$ or $j < k - r + 2$). By construction, for all i,

$$\mu_i^{(r)} = \mu_i^{(1)} + d_{i+1} - d_i$$

and

$$p^{\ell_p(d_i)} \text{ divides } \mu_{i-1}^{(1)} - \mu_i^{(1)} - d_{i+1} + d_i + 1$$

Let $N_{i1} = \mu_i^{(1)} - \mu_{i+1}^{(r)}$ and $N_{ij} = \mu_{i+j-1}^{(r)} - \mu_{i+j}^{(r)}$ for $j \geq 2$, and let T be the corresponding $\mu^{(r)}$-tableau of type $\mu^{(1)}$ in Theorem 24.6 (It is simple to verify that T is reverse semistandard).

Now,
$$\sum_{m=1}^{i-1} \left(\mu_{j+m-1}^{(r)} - \sum_{s=j}^{\infty} N_{ms}\right) = d_i \quad \text{if } j = 1, \text{ and } 0 \quad \text{if } j \geq 2.$$

Also, $N_{i-1,1} = \mu_{i-1}^{(1)} - \mu_i^{(r)} = \mu_{i-1}^{(1)} - \mu_i^{(1)} - d_{i+1} + d_i \equiv -1 \mod p^{\ell_p(d_i)}$, so Theorem 24.6(ii) gives the result.

24.11 EXAMPLE Suppose char $F = 3$

$$\begin{array}{lclclcl}
\cdot\ \cdot\ \cdot\ \cdot\ \cdot & \overset{4}{\to} & \cdot\ \cdot\ \cdot\ \cdot\ \cdot & \overset{3}{\to} & \cdot\ \cdot\ \cdot\ \cdot\ \cdot & \overset{2}{\to} & \cdot\ \cdot\ \cdot\ \cdot\ \cdot\ \cdot\ \cdot \\
\cdot\ \cdot\ \cdot & & \cdot\ \cdot\ \cdot & & \cdot\ \cdot\ \cdot\ \times\ \times & & \cdot\ \cdot\ \cdot \\
\cdot\ \cdot & & \cdot\ \times\ \times & & \cdot & & \cdot \\
\times & & & & & &
\end{array}$$

Therefore, $\mathrm{Hom}_{F\mathfrak{S}_{11}}(S^{\lambda}, S^{\mu}) \neq 0$ for $\lambda \trianglerighteq \mu$ and λ, μ any pair from (7,3,1),

$(5^2,1)$, $(5,3^2)$ and $(5,3,2,1)$. Compare the following 4 by 4 submatrix of the decomposition matrix of $\mathfrak{S}_{11}$ for the prime 3.

	$D^{(7,3,1)}$	$D^{(5^2,1)}$	$D^{(5,3^2)}$	$D^{(5,3,2,1)}$
$S^{(7,3,1)}$	1			
$S^{(5^2,1)}$	1	1		
$S^{(5,3^2)}$	1	1	1	
$S^{(5,3,2,1)}$	1	1	1	1

Note that the number of nodes we raise to the row above need not be the same for each $\mu^{(k-j+1)} \overset{j}{\to} \mu^{(k-j+2)}$ in Corollary 24.10; in particular, the Corollary includes the case

$$\mu^{(1)} \overset{i_1}{\to} \mu^{(2)} \overset{i_2}{\to} \mu^{(3)} \to \dots \overset{i_{r-1}}{\to} \mu^{(r)} \quad \text{with } i_1 > i_2 > \dots > i_{r-1}$$

since we are allowed to raise zero nodes at any stage. The hypothesis $i_1 > i_2 > \dots > i_{r-1}$ cannot be omitted, since when char F = 2,

$$\begin{matrix} X & X \\ X & X \end{matrix} \overset{2}{\to} \begin{matrix} X & X & X \\ X & & \end{matrix} \overset{2}{\to} \begin{matrix} X & X & X & X \end{matrix}$$

and while $\operatorname{Hom}_{F\mathfrak{S}_4}(S^{(4)}, S^{(3,1)})$ and $\operatorname{Hom}_{F\mathfrak{S}_4}(S^{(3,1)}, S^{(2^2)})$ are non-zero (by the Corollary), $\operatorname{Hom}_{F\mathfrak{S}_4}(S^{(4)}, S^{(2^2)})$ is zero (by Theorem 24.4).

For our next Theorem we require

24.12 DEFINITION Given two non-negative integers a and b, let

$$a = a_0 + a_1 p + \dots + a_r p^r \quad (0 \le a_i < p,\ a_r \ne 0)$$

$$b = b_0 + b_1 p + \dots + b_s p^s \quad (0 \le b_i < p,\ b_s \ne 0).$$

We say that <u>a contains b to base p</u> if $s < r$ and for each i $b_i = 0$ or $b_i = a_i$.

24.13 EXAMPLE $65 = 2 + 0.3 + 1.3^2 + 2.3^3$, so 65 contains precisely $0, 2, 9 = 1.3^2$ and $11 = 2 + 1.3^2$ to base 3.

24.14 DEFINITION The function $f_p(n,m)$ is defined by $f_p(n,m) = 1$ if $n+1$ contains m to base p, and $= 0$, otherwise.

Since the only composition factors of $S^{(n-m,m)}$ have the form $D^{(n-j,j)}$ with $j \le m$, by Corollary 12.2, a sensible first step towards evaluating the decomposition matrix for $\mathfrak{S}_n$ is to prove

24.15 THEOREM (James [6] and [8]). <u>The multiplicity of $D^{(n-j,j)}$ as a factor of $S^{(n-m,m)}$ is $f_p(n-2j, m-j)$.</u>

<u>Proof</u> Since the result is true when n = 0 or 1, we may assume it for

$n' < n$. Let t be the $(n-j,j)$-tableau used to define the $\mathfrak{S}_n$ action on $M^{(n-m,m)}$. Let T be the $(n-j,j)$-tableau of type $(n-m,m)$ having 2's in the $(1,1)$th, $(1,2)$th, ..., $(1,m)$th places. As in the proof of Theorem 24.6, the ψ maps defined on $M^{(n-m,m)}$ have the property that

$$\{t\}\theta_T \in \bigcap_{i=r}^{m-1} \ker \psi_{1,i} \quad \text{if } n-m-j \equiv -1 \bmod p^{\ell_p(m-r)}.$$

Also

$$\ker \theta_T \subseteq S^{(n-j,j)\perp} .$$

Therefore, all the composition factors of $S^{(n-j,j)}$ occur in $\bigcap_{i=r}^{m-1} \ker \psi_{1,i}$

But, by the second isomorphism theorem,

$$\bigcap_{i=r}^{m-1} \ker \psi_{1,i} \Big/ \bigcap_{i=o}^{m-1} \ker \psi_{1,i} \cong \Big(\bigcap_{i=r}^{m-1} \ker \psi_{1,i} + \bigcap_{i=o}^{r-1} \ker \psi_{i,r}\Big) \Big/ \bigcap_{i=o}^{r-1} \ker \psi_{1,i}$$

$$\subseteq M^{(n-m,m)} \Big/ \bigcap_{i=o}^{r-1} \ker \psi_{1,i} .$$

Thus, every composition factor of $\bigcap_{i=r}^{m-1} \ker \psi_{1,i}$ is either a factor of $S^{(n-m,m)} = \bigcap_{i=o}^{m-1} \ker \psi_{1,i}$ or of $M^{(n-m,m)} / \bigcap_{i=o}^{r-1} \ker \psi_{1,i}$. By Theorem 17.13 we have:

<u>24.16</u> If $n-m-j \equiv -1 \bmod p^{\ell_p(m-r)}$, then every factor of $S^{(n-j,j)}$ is a factor of $S^{(n-m,m)}$ or of one of $\{S^{(n-i,i)} \mid 0 \le i \le r-1\}$.

Now suppose that $f_p(n-2j,m-j) = 1$. Then $m \ge j \ge 0$ and $n-2j+1$ contains $m-j$ to base p. If $m > j$, then there is a unique integer j_1 such that

$$n-2j+1 \equiv (m-j) + (j_1-j) \bmod p^{\ell_p(m-j)}$$

and $\quad 0 \le j_1-j < m-j$.

But then $n-2j+1$ contains j_1-j to base p. Hence we may find integers such that

$$m = j_o > j_1 > \dots \; j_s > j_{s+1} = j$$

and $\quad n - j_k - j_{k+1} \equiv -1 \bmod p^{\ell_p(j_k-j)}$.

Then, by 24.16 every factor of $S^{(n-j,j)}$ is a factor of $S^{(n-j_s,j_s)}$ or one of $\{S^{(n-i,i)} \mid 0 \le i \le j-1\}$. But $D^{(n-j,j)}$ is not a factor of $S^{(n-i,i)}$ for $0 \le i \le j-1$, by Corollary 12.2, so $D^{(n-j,j)}$ is a factor of $S^{(n-j_s,j_s)}$.

Applying 24.16 again, every factor $S^{(n-j_s,j_s)}$ is a factor of

$S^{(n-j_{s-1},j_{s-1})}$ or of one of $\{S^{(n-i,i)} | 0 \le i \le j-1\}$. Therefore, $D^{(n-j,j)}$ is a factor of $S^{(n-j_{s-1},j_{s-1})}$. Continuing this argument to $j_o = m$, we have proved

<u>24.17</u> When $f_p(n-2j,m-j) = 1$, $D^{(n-j,j)}$ is a factor of $S^{(n-m,m)}$.

Next, consider the case where $n \equiv m-1 \bmod p^{\ell_p(m)}$. Then let

$$m-1 = a_o + a_1 p + \ldots + a_{r-1}p^{r-1} \quad (0 \le a_i < p,\ a_{r-1} \ne 0)$$

so $$n = a_o + a_1 p + \ldots + a_{r-1}p^{r-1} + b_r p^r + \ldots$$

where $b_r = 0$ if $m = p^r$. Thus, n contains m-1 to base p, so $f_p(n-1,m-1) = 1$. Similarly, $f_p(n-1,m) = 0$ and $f_p(n,m) = 1$.

Returning to the case of general n and m, we prove

<u>24.18</u> If $m \ge 1$ and $f_p(n-1,m) + f_p(n-1,m-1) > f_p(n,m)$, then there is some integer j with $1 \le j \le m$ such that $D^{(n-j,j)}$ is a factor of $S^{(n-m,m)}$ and $D^{(n-j,j)} \downarrow \mathfrak{S}_{n-1}$ contains the trivial factor $D^{(n-1)}$ with multiplicity $f_p(n-1,m) + f_p(n-1,m-1)$.

To prove 24.18, consider first the case where m is a power of p, say $m = p^r$. The inequality $f_p(n-1,m) + f_p(n-1,m-1) > f_p(n,m)$ easily implies that p^r divides $n + 1$, and the argument above proves that p^{r+1} does not divide n-m+1. Therefore, $\nu_p(n-m+1) = r$. Hence $S^{(n-m,m)}$ is irreducible in this case, by Theorem 23.13, and $D^{(n-m,m)} = S^{(n-m,m)}$. Since $S^{(n-m,m)} \downarrow \mathfrak{S}_{n-1}$ has the same factors as $S^{(n-m-1,m)} \oplus S^{(n-m,m-1)}$ by the Branching Theorem, $D^{(n-m,m)} \downarrow \mathfrak{S}_{n-1}$ contains $D^{(n-1)}$ with multiplicity $f_p(n-1,m) + f_p(n-1,m-1)$, by the induction hypothesis. This shows that we may take $j = m$ in 24.18 when m is a power of p.

Suppose, therefore, that m is not a power of p. Since $f_p(n-1,m) + f_p(n-1,m-1) \ge 1$, n contains m or m-1 to base p. The fact that m is not a power of p now shows there is a unique j with

$$0 \le j < m \qquad n \equiv m+j-1 \bmod p^{\ell_p(m)}.$$

Further, $j \ge 1$, since we have shown that $n \equiv m-1 \bmod p^{\ell_p(m)}$ implies that $f_p(n-1,m) + f_p(n-1,m-1) = f_p(n,m)$. Now the above congruence shows that n + 1 contains m to base p if and only if n+1 contains j to base p, and n contains m to base p if and only if n contains j-1 to base p, and n contains m-1 to base p if and only if n contains j to base p. Therefore

$$f_p(n-1,j) + f_p(n-1,j-1) = f_p(n-1,m) + f_p(n-1,m-1)$$
$$> f_p(n,m) = f_p(n,j).$$

By induction, there is an i with $1 \le i \le j < m$ such that $D^{(n-i,i)}$ is a factor of $S^{(n-j,j)}$ and $D^{(n-i,i)} \downarrow \mathfrak{S}_{n-1}$ has $D^{(n-1)}$ as a factor with multiplicity $f_p(n-1,m) + f_p(n-1,m-1)$. But, since $n \equiv m+j-1 \bmod p^{\ell}p^{(m)}$, 24.16 shows that every factor of $S^{(n-j,j)}$ is a factor of $S^{(n-m,m)}$. In particular, $D^{(n-i,i)}$ is a factor of $S^{(n-m,m)}$ and so 24.18 is proved.

The multiplicity of $D^{(n)}$ as a factor of $S^{(n-m,m)}$ is at most $f_p(n-1,m) + f_p(n-1,m-1)$, since $S^{(n-m,m)}\downarrow \mathfrak{S}_{n-1}$ has $D^{(n-1)}$ as a factor with this multiplicity, by our induction hypothesis. Further, 24.18 shows that $D^{(n)}$ is not a factor of $S^{(n-m,m)}$ when $f_p(n-1,m) + f_p(n-1,m-1) > f_p(n,m)$. This proves our next main result, namely

<u>24.19</u> The multiplicity of $D^{(n)}$ as a factor of $S^{(n-m,m)}$ is at most $f_p(n,m)$.

Finally we prove

<u>24.20</u> If $j \geq 1$, $D^{(n-j,j)}$ is a factor of $S^{(n-m,m)}$ with multiplicity at most $f_p(n-2j,m-j)$.

The way we show this is to consider a subgroup H of $\mathfrak{S}_n$, and find a modular representation D_j of H such that $D^{(n-j,j)}\downarrow H$ has D_j as a factor, but $S^{(n-m,m)}\downarrow H$ has D_j as a factor with multiplicity $f_p(n-2j, m-j)$. 24.20 then follows at once. We should like to choose $\mathfrak{S}_{n-2}$ or $\mathfrak{S}_{n-1}$ as our subgroup H, so that we can apply induction. Since the prime 2 is exceptional, we consider first

<u>Case 1</u> p is odd.

The ordinary irreducible representations of $\mathfrak{S}_{(n-2,2)}$ are given by $S^{\mu}_{\mathbb{Q}} \otimes S^{(2)}_{\mathbb{Q}}$ and $S^{\mu}_{\mathbb{Q}} \otimes S^{(1^2)}_{\mathbb{Q}}$ as μ varies over partitions of n-2. Since p is odd, $D^{(2)}$ and $D^{(1^2)}$ are inequivalent representations. Hence the p-modular irreducible representations of $\mathfrak{S}_{(n-2,2)}$ are given by $D^{\mu} \otimes D^{(2)}$, $D^{\mu} \otimes D^{(1^2)}$ as μ varies over p-regular partitions of n-2, and the multiplicity of $D^{(n-j-1,j-1)} \otimes D^{(1^2)}$ as a factor of $S^{(n-m-1,m-1)} \otimes S^{(1^2)}$ is $f_p(n-2j,m-j)$ when $j \geq 1$, by induction.

Now, by the Littlewood-Richardson Rule, $S^{(n-m,m)}\downarrow \mathfrak{S}_{(n-2,2)}$ has the same composition factors as $S^{(n-m-1,m-1)} \otimes S^{(1^2)}$, together with some modules of the form $S^{\mu} \otimes S^{(2)}$. In particular, the multiplicity of $D^{(n-j-1,j-1)} \otimes D^{(1^2)}$ as a factor of $S^{(n-m,m)}\downarrow \mathfrak{S}_{(n-2,2)}$ is $f_p(n-2j,m-j)$.

On the other hand, $S^{(n-j,j)}\downarrow \mathfrak{S}_{(n-2,2)}$ has $D^{(n-j-1,j-1)} \otimes D^{(1^2)}$ as a factor with multiplicity one (since $f_p(n-2j,0) = 1$), and for $i < j$ $S^{(n-i,i)}\downarrow \mathfrak{S}_{(n-2,2)}$ does not have $D^{(n-j-1,j-1)} \otimes D^{(1^2)}$ as a factor (since $f_p(n-2j,i-j) = 0$). Now, every factor of $S^{(n-j,j)}$, besides $D^{(n-j,j)}$, has the form $D^{(n-i,i)}$ with $i < j$, so it follows that $D^{(n-j,j)}\downarrow \mathfrak{S}_{(n-2,2)}$ has $D^{(n-j-1,j-1)} \otimes D^{(1^2)}$ as a factor with multiplicity one.

The results of the last two paragraphs prove 24.20 in this case.

<u>Case 2a</u> $p = 2$ and n is even.

$S^{(n-m,m)} \downarrow \mathfrak{S}_{n-1}$ has the same factors as $S^{(n-m-1,m)} \oplus S^{(n-m,m-1)}$. By induction, this contains the factor $D^{(n-j-1,j)}$ with multiplicity $f_2(n-1-2j,m-j) + f_2(n-1-2j,m-j-1)$. It is simple to verify that this equals $f_2(n-2j,m-j)$, since n is even.

In particular, for $2j < n$, $S^{(n-j,j)} \downarrow \mathfrak{S}_{n-1}$ has $D^{(n-j-1,j)}$ as a factor with multiplicity one, and for $i < j$, $S^{(n-i,i)} \downarrow \mathfrak{S}_{n-1}$ does not have $D^{(n-j-1,j)}$ as a factor. As before, $D^{(n-j,j)} \downarrow \mathfrak{S}_{n-1}$ therefore has $D^{(n-j-1,j)}$ as a factor with multiplicity one, and 24.20 is proved in this case too.

<u>Case 2b</u> $p = 2$ and n is odd.

$S^{(n-m,m)} \downarrow \mathfrak{S}_{n-2}$ has the same factors as $S^{(n-m-2,m)} \oplus 2\, S^{(n-m-1,m-1)} \oplus S^{(n-m,m-2)}$. This contains $D^{(n-j-1,j-1)}$ with multiplicity $f_2(n-2j,m-j+1) + 2f_2(n-2j,m-j) + f_2(n-2j,m-j-1)$, which equals $2f_2(n-2j, m-j)$ when $m-j$ is even,

Thus, $S^{(n-j,j)} \downarrow \mathfrak{S}_{n-2}$ has $D^{(n-j-1,j-1)}$ as a factor with multiplicity 2, and for $i \le j-2$, $S^{(n-i,i)} \downarrow \mathfrak{S}_{n-2}$ does not have $D^{(n-j-1,j-1)}$ as a factor. But every factor of $S^{(n-j,j)}$, besides $D^{(n-j,j)}$, has the form $D^{(n-i,i)}$ with $i \le j-2$, by the Remark following Theorem 23.7, so $D^{(n-j,j)} \downarrow \mathfrak{S}_{n-2}$ has $D^{(n-j-1,j-1)}$ as a factor with multiplicity 2.

The results of the last two paragraphs prove 24.20 in this final case.

Now 24.17, 24.19 and 24.20 together give Theorem 24.15.

24.21 COROLLARY <u>If $j \ge 1$, the multiplicity of $D^{(n-j,j)}$ as a factor of $S^{(n-m,m)}$ is the same as the multiplicity of $D^{(n-j-1,j-1)}$ as a factor of $S^{(n-m-1,m-1)}$</u>.

By the way, we conjecture that Corollary 24.21 is a special case of a general theorem involving the removal of the first column.

24.22 EXAMPLE Suppose $p = 3$. The rows of the following table record, respectively, n, $n+1$ written to base 3, and the numbers contained in $n+1$ to base 3, for $0 \le n \le 13$.

0	1	2	3	4	5	6	7	8	9	10	11	12	13
1	2	10	11	12	20	21	22	100	101	102	110	111	11
0	0	0	0	0	0	0	0	0	0	0	0	0	0
			1	2		1	2		1	2	10	1	2
												10	10
												11	12

Under n = 13, for example, we have 0,2,10,12 which are integers to base 3. There are 1's in the (0+1)th, (2+1)th, (3+1)th and (5+1)th places (counting from the diagonal) in the column labelled 13 in the following pair of matrices. Another example: 10+1 contains 0 and 2 to base 3. There are 1's in the (0+1)th and (2+1)th places of the column labelled 10.

12	10	8	6	4	2	0
1						
1	1					
		1				
1	1		1			
1			1	1		
					1	
				1		1

13	11	9	7	5	3	1
1						
	1					
1		1				
1		1	1			
	1			1		
1			1		1	
					1	1

The part of the decomposition matrix of $\mathfrak{S}_n$ corresponding to 2-part partitions for p = 3 and n ≤ 13 can be read off these matrices at once. Simply truncate the matrix at the column labelled n, and label the rows and columns by 2-part partitions in dictionary order.

e.g. n = 9

	(9)	(8,1)	(7,2)	(6,3)	(5,4)
(9)	1				
(8,1)	1	1			
(7,2)			1		
(6,3)		1		1	
(5,4)				1	1

For p an odd prime and n small, most of the decomposition matrix of $\mathfrak{S}_n$ is given by Theorems 24.1 and 24.15.

24.33 EXAMPLE Suppose p = 3 and n = 9. Applying Peel's Theorem 24.1, the column labels can be found as in Example 24.2 . Alternatively, they are given explicitly in [9] page 52. Combined with the information above, this gives

	(9)	(8,1)	(7,2)	(6,3)	(5,4)	$(7,1^2)$	(6,2,1)	$(5,2^2)$	(4,3,2)	$(4^2,1)$
(9)	1									
(8,1)	1	1								
(7,2)			1							
(6,3)		1		1						
(5,4)				1	1					
$(7,1^2)$		1				1				
$(6,1^3)$						1	1			
$(5,1^4)$							1	1		

$(4,1^5)$		1	1	
$(3,1^6)$			1	1
$(2,1^7)$	1			1
(1^9)	1			

Applying Theorem 8.15 to the first five rows, another part of the decomposition matrix is

	$(5,4)$	$(4^2,1)$			
(1^9)	1				
$(2,1^7)$	1	1			
$(2^2,1^5)$			1		
$(2^3,1^3)$		1		1	
$(2^4,1)$				1	1

(The rows corresponding to (1^9) and $(2,1^7)$ already occur above). Using Theorem 21.7 we find that the last three columns should be labelle $(4,3,1^2)$, $(3^2,2,1)$ and (9). Incidentally, we do not know how to sort out efficiently the column labels once we have taken conjugate partitions as above (although Theorem A in [9] gives some partial answers).

We have now accounted for 12 of the 16 3-regular partitions labelling columns. $S^{(5,3,1)}$ and $S^{(3,2^2,1)}$ are irreducible, by Example 23.6(i), so we have two more 3-modular irreducibles to find, namely those corresponding to $(4,2^2,1)$ and $(5,2,1^2)$. But

$$\chi^{(7,2)} - \chi^{(4,2^2,1)} + \chi^{(4,2,1^3)}$$

on 3-regular classes (using Theorem 21.7 with $[\nu] = [4,2]$). Appealing to the theory of blocks of defect 1 (or to the Nakayama Conjecture) part of our decomposition matrix is

	$(7,2)$	$(4,2^2,1)$
$(7,2)$	1	
$(4,2^2,1)$	1	1
$(4,2,1^3)$		1

By taking conjugate partitions, we get

	$(5,2,1^2)$	$(4,3,1^2)$
$(5,2,1^2)$	1	
$(4,3,1^2)$	1	1
$(2^2,1^5)$		1

Now Theorem 21.7 enables us to complete the decomposition matrix, since we can write every ordinary character which corresponds to a 3-singular partition in terms of ordinary characters corresponding to 3-regular partitions, on 3-regular classes.

When p = 2, Theorem 24.1 cannot be applied. However, all the rows of the decomposition matrix for partitions of the form (n-m-1,m,1) are known for p = 2 (see James [6]).

Our sources for the decomposition matrices in the Appendix are Kerber [13] ($p = 2, n \leq 9$), James [6]($p = 2$, $n = 10$), Mac Aogáin [15] ($p = 2, n = 11$), Stockhofe [21] ($p = 2, n = 12, 13$), Kerber and Peel [14] ($p = 3$, $8 \leq n \leq 10$) and Mac Aogáin [15]($p = 3, 11 \leq n \leq 13$, completed by James [12]). Mac Aogáin [15] gives the decomposition matrices for $p = 5, n \leq 13$.

The most difficult cases are $p = 2, n = 12$ and 13, and for these Stockhofe used a computer to find dim $D^{(5,4,2,1)}$ and dim $D^{(7,4,2)}$, employing Theorem 11.6.

25 YOUNG'S ORTHOGONAL FORM

We turn now to the problem of finding the matrices which represent the action of permutations on the Specht module S^μ. This has been postponed to a late stage in order to emphasize the fact that the representation theory of $\mathfrak{S}_n$ can (and we believe <u>should</u>) be presented without reference to the representing matrices.

Since $\mathfrak{S}_n$ is generated by the transpositions $(x-1,x)$ for $1 < x \le n$ is is sufficient to determine the action of these transposition on a basis of S^μ. Consider first the basis of standard polytabloids e_t. Here we have

<u>25.1</u> (i) If $x-1$ and x are in the same column of t, then $e_t(x-1,x) = -e_t$

(ii) If $x-1$ and x are in the same row of t, then $e_t(x-1,x) = e_t$ + a linear combination of standard polytabloids e_{t^*} with $\{t^*\} \lhd \{t\}$ (by combining 8.3 and the technique used to prove 8.9).

(iii) If $t(x-1,x)$ is standard, then $e_t(x-1,x) = e_{t(x-1,x)}$.

In case (ii), the relevant standard tableaux t^* may be calculated by applying the Garnir relations.

25.2 EXAMPLE If $\mu = (3,2)$ and we take the standard μ-tableau in the order $t_1,t_2,t_3,t_4,t_5 = \begin{matrix}1&3&5\\2&4\end{matrix} \quad \begin{matrix}1&2&5\\3&4\end{matrix} \quad \begin{matrix}1&3&4\\2&5\end{matrix} \quad \begin{matrix}1&2&4\\3&5\end{matrix} \quad \begin{matrix}1&2&3\\4&5\end{matrix}$ then

$$(1\ 2) \leftrightarrow \begin{pmatrix}-1&0&0&0&0\\-1&1&0&0&0\\0&0&-1&0&0\\0&0&-1&1&0\\1&0&-1&0&1\end{pmatrix} \qquad (2\ 3) \leftrightarrow \begin{pmatrix}0&1&0&0&0\\1&0&0&0&0\\0&0&0&1&0\\0&0&1&0&0\\1&-1&0&0&1\end{pmatrix}$$

$$(3\ 4) \leftrightarrow \begin{pmatrix}-1&0&0&0&0\\-1&1&0&0&0\\-1&0&1&0&0\\0&0&0&0&1\\0&0&0&1&0\end{pmatrix} \qquad (4\ 5) \leftrightarrow \begin{pmatrix}0&0&1&0&0\\0&0&0&1&0\\1&0&0&0&0\\0&1&0&0&0\\1&0&-1&0&1\end{pmatrix}$$

In many ways, Young's natural representation, as this is called, is the best way of describing the matrices which represent permutations; for example, it is independent of the field. However, we must take three cases into account, and the second one, where $x-1$ and x are in the same row, involves an unpleasant calculation. It turns out that these problems can be avoided when we work over the field $\mathbb{R}$ of real numbers, and the rest of this section will be devoted to the case where <u>the ground</u>

field is $\mathbb{R}$.

Let $t_1 < t_2 < \ldots < t_d$ be the standard μ-tableaux, in the order given by definition 3.10. Wherever possible, we shall use the abbreviation e_i for the standard polytabloid e_{t_i}.

Since we are working over the reals, we may construct from $e_1, e_2, \ldots, e_d$ an orthonormal basis $f_1, f_2, \ldots, f_d$ of $S^\mu_{\mathbb{R}}$ using the Gram-Schmidt orthogonalization process. It is with respect to the new orthonormal basis that we get "nice" matrices representing permutations. To fix notation, we remind the reader of the Gram-Schmidt orthogonalization process.

Suppose we have constructed a basis $f_1, \ldots, f_j$ of the space spanned by $e_1, \ldots, e_j$ over $\mathbb{R}$, and that $f_1, \ldots, f_j$ are orthonormal relative to the bilinear form $\langle \ , \ \rangle$. Then there is a non-zero linear combination f of $e_1, \ldots, e_{j+1}$ with $\langle e_i, f \rangle = 0$ for $1 \le i \le j$ (see 1.3). Now, the tabloid $\{t_{j+1}\}$ is involved in f (otherwise f would be a linear combination of $e_1, \ldots, e_j$ by the proof of 8.9, contradicting the fact that $\langle e_i, f \rangle = 0$ for $1 \le i \le j$.) Therefore, we may take

$$f_{j+1} = (\pm f)/(\langle f, f \rangle)^{\frac{1}{2}},$$

the sign being chosen so that $\{t_{j+1}\}$ has a positive coefficient in f_{j+1}. This determines f_{j+1} uniquely.

Of course, the new basis $f_1, f_2, \ldots, f_d$ of $S^\mu_{\mathbb{R}}$ depends on the order of the original basis $e_1, e_2, \ldots, e_d$. However, we prove

25.3 THEOREM <u>The orthonormal basis $f_1, f_2, \ldots, f_d$ of $S^\mu_{\mathbb{R}}$ constructed from the standard basis is independent of the total order we choose on the standard tableaux, provided that the total order contains the partial order $\lhd$, given by definition 3.11</u>.

At the same time, we prove

25.4 YOUNG'S ORTHOGONAL FORM

<u>If $(x-1,x)$ is a transposition in $\mathfrak{S}_n$, then for all r</u>

$$f_r(x-1,x) = \rho_1 f_r + \rho_2 f_s$$

<u>where $t_s = t_r(x-1,x)$ and $\rho_1 (= \rho_1(x,r))$ equals $(i-k+\ell-j)^{-1}$ if $x-1$ is in the (i,j)th position and x is in the (k,ℓ)th position of t_r, and $\rho_1^2 + \rho_2^2 = 1$ with $\rho_2 \ge 0$.</u>

<u>Remark</u>: It does not matter that there is no t_s equal to $t_r(x-1,x)$ when $x-1$ and x are in the same row or column of t_r, since $\rho_2 = 0$ in these cases. Young's Orthogonal Form says that $f_r(x-1,x) = \pm f_r$ if $x-1$ and x are in the same row or column of t_r, respectively.

Before embarking on the proofs of 25.3 and 25.4, we require a preliminary Lemma.

25.5 LEMMA <u>Suppose that t and t^* are any two μ-tableaux, and that $x-1$ is lower than x in t^*. If $\{t\} \lhd \{t^*\}$ then $\{t\}(x-1,x) \lhd \{t^*\}(x-1,x)$.</u>

<u>Proof</u>: Recall from definition 3.11 that $m_{iu}(t)$ is the number of entries less than or equal to i in the first u rows of t. Since $\{t\} \lhd \{t^*\}$, $m_{iu}(t) \le m_{iu}(t^*)$ for all i and u.

Let $x-1$ be in the a_1th row and x be in the b_1th row of t. Let $x-1$ be in the a_2th row and x be in the b_2th row of t^*. We are given that $b_2 < a_2$.

Using 3.14, we deduce from $m_{iu}(t) \le m_{iu}(t^*)$ that $m_{iu}(t(x-1,x)) \le m_{iu}(t^*(x-1,x))$, except perhaps for $i = x-1$ and either $b_1 \le u < \min(a_1,b_2)$ or $\max\ (b_1,a_2) \le u < a_1$.

For $b_1 \le u < \min(a_1,b_2)$,

$$\begin{aligned} m_{x-1,u}(t(x-1,x)) &= m_{x,u}(t), \text{ since } x-1 \text{ is in the } a_1\text{th row and } x \text{ is in the } b_1\text{th row of } t \text{ and } b_1 \le u < a_1 \\ &\le m_{x,u}(t^*), \text{ since } \{t\} \lhd \{t^*\} \\ &= m_{x-1,u}(t^*(x-1,x)), \text{ since } u < b_2 < a_2 . \end{aligned}$$

For $\max\ (b_1,a_2) \le u < a_1$,

$$\begin{aligned} m_{x-1,u}(t(x-1,x)) &= m_{x-2,u}(t) + 1, \text{ since } b_1 \le u < a_1 \\ &\le m_{x-2,u}(t^*) + 1, \text{ since } \{t\} \lhd \{t^*\} \\ &= m_{x-1,u}(t^*(x-1,x)), \text{ since } b_2 < a_2 \le u . \end{aligned}$$

Therefore, $m_{iu}(t(x-1,x)) \le m_{iu}(t^*(x-1,x))$ in all cases. Thus $\{t(x-1,x)\} \unlhd \{t^*(x-1,x)\}$. We do not have equality, since $\{t\} \ne \{t^*\}$.

<u>Proofs of Theorem 25.3 and Young's Orthogonal Form</u>:

Assume that both results are true for all $\mathbb{R}\mathfrak{S}_{n-1}$ Specht modules (Both are vacuously true when $n = 0$). The proof now proceeds in 3 steps.

<u>Step 1</u> The matrices which we claim represent $(x-1,x)$ are correct for $x < n$.

We take our notation for the proof of Theorem 9.3, so that V_i is the $\mathbb{R}\mathfrak{S}_{n-1}$-module spanned by those e_t's where t is a standard μ-tableau, and n is in the r_ith, r_2th,...,or r_ith row of t. Since $V_1 \subset V_2 \subset \ldots$, the proof we gave for Maschke's Theorem shows that

$$V_i = U_1 \oplus U_2 \oplus \ldots \oplus U_i,$$

where U_i is the $\mathbb{R}\mathfrak{S}_{n-1}$-module spanned by those f_t's where n is in the

r_ith row of t. (Recall that our total order on tabloids puts all those with n in the r_1th row before all those with n in the r_2th row etc.)

In the proof of Theorem 9.3 we constructed an $\mathbb{R}\mathfrak{S}_{n-1}$-homomorphism θ_i mapping V_i onto $S^{\lambda^i}_{\mathbb{R}}$ whose kernel is V_{i-1}. Since $V_{i-1} = U_1 \oplus \ldots \oplus U_{i-1}$ and $V_i = U_1 \oplus \ldots \oplus U_i$, we therefore know that θ_i is an $\mathbb{R}\mathfrak{S}_{n-1}$-isomorphism from U_i onto $S^{\lambda^i}_{\mathbb{R}}$.

Define a bilinear form $< \ , \ >^*$ on U_i by

$$< u,v >^* = < u\theta_i , v\theta_i > \quad \text{for } u,v \text{ in } U_i ,$$

where the second bilinear form is that on $S^{\lambda^i}_{\mathbb{R}}$. Since U_i is an absolutely irreducible $\mathbb{R}\mathfrak{S}_{n-1}$-module, our new bilinear form on U_i must be a multiple of the original one, by Schur's Lemma. That is, there is a real constant c such that

$$< u,v >^* = c< u,v > \text{ for all } u,v \text{ in } U_i .$$

Because both forms are inner products, c is positive.

For each standard μ-tableau t having n in the r_ith row, let $\bar{t}$ denote t with n removed, and write $\bar{e}_t$ for $e_{\bar{t}}$ and $\bar{f}_t$ for $f_{\bar{t}}$. Suppose that $t_p, t_{p+1}, \ldots, t_q$ are the standard μ-tableaux which have n in the r_ith row. If $p \le r \le q$ then

$$f_r = u + a_p e_p + a_{p+1} e_{p+1} + \ldots + a_r e_r$$

for some u in V_{i-1} and $a_r > 0$. Therefore, by 9.4,

$$f_r\theta_i = a_p\bar{e}_p + a_{p+1}\bar{e}_{p+1} + \ldots + a_r\bar{e}_r .$$

Since the last tabloid here is $\{\bar{t}_r\}$ with a positive coefficient, and since $< f_z\theta_i , f_r\theta_i > = c< f_z , f_r >$ for $p \le z \le r$, we deduce that

$$f_r \theta_i = \sqrt{c}\, \bar{f}_r .$$

We are assuming that Young's Orthogonal Form is correct for the $\mathbb{R}\mathfrak{S}_{n-1}$-module S^{λ^i}, so for $x < n$,

$$\begin{aligned} f_r(x-1,x)\theta_i &= \sqrt{c}\, \bar{f}_r(x-1,x) \\ &= \sqrt{c}\, (\rho_1\bar{f}_r + \rho_2\bar{f}_s) = (\rho_1 f_r + \rho_2 f_s)\theta_i . \end{aligned}$$

Here, $t_s = t_r(x-1,1)$, and the real numbers ρ_1 and ρ_2 are those in the statement of Young's Orthogonal Form (the positions of x-1 and x in t_r are the same as their positions in $\bar{t}_r$). Since θ_i is an isomorphism, we have proved the desired result of Step 1, namely that

$$f_r(x-1,x) = \rho_1 f_r + \rho_2 f_s, \text{ for } x < n.$$

<u>Step 2</u> The proof of Theorem 25.3 .

We know that there are real numbers $a_1, a_2, \ldots, a_r$ with

$$f_r = a_1e_1 + a_2e_2 + \ldots + a_re_r \quad \text{and } a_r > 0.$$

Theorem 25.3 will follow if we can show that $a_j = 0$ unless $\{t_j\} \trianglelefteq \{t_r\}$. By induction, we may assume that when $\{t_j\} \triangleleft \{t_r\}$, f_j is a linear combination of standard polytabloids e_i with $\{t_i\} \trianglelefteq \{t_j\}$, and prove the corresponding result for f_r.

Case 1 For some $x < n$, x is lower than $x-1$ in t_r and not in the same row or column as $x-1$.

Let $t_r(x-1,x) = t_k$. Then $\{t_k\} \triangleleft \{t_r\}$. Therefore,

$$f_k = c_1e_1 + \ldots + c_ke_k \quad \text{where } c_i = 0 \text{ unless } \{t_i\} \trianglelefteq \{t_k\}.$$

Using 25.1, and applying Lemma 25.5, $f_k(x-1,x)$ is a linear combination of polytabloids e_i for which $\{t_i\} \trianglelefteq \{t_r\}$.

Since $x < n$, Step 1 shows that

$$f_r = \text{a multiple of } f_k + \text{a multiple of } f_k(x-1,x).$$

Therefore in this case,

$$f_r = a_1e_1 + \ldots + a_re_r \quad \text{where } a_j = 0 \text{ unless } \{t_j\} \trianglelefteq \{t_r\}.$$

Case 2 For every $x < n$, x is higher than $x-1$ in t_r or is in the same row or column as $x-1$.

Since t_r is standard, it is easy to see that the hypothesis of Case 2 implies that $\bar{t}_r$ (= t_r, with n removed) has $1,2,\ldots,n-1$ in order down successive columns.

We may certainly write

$$f_r = b_1f_1 + \ldots + b_{r-1}f_{r-1} + b_re_r \quad \text{where } b_r \neq 0.$$

Let x be the smallest integer such that $b_j \neq 0$ for some j and $m_{xu}(t_r) < m_{xu}(t_j)$ for some u, if such an integer x exists. We aim to produce a contradiction.

First, $1 < x < n$, since for all u, $m_{1u}(t_r) = m_{1u}(t_j) = 1$ (t_r and t_j being standard), and $m_{nu}(t_r) = m_{nu}(t_j) = \mu_1 + \ldots + \mu_u$ for all μ-tableaux t_r and t_j.

By the minimality of x, $m_{x-1,u}(t_r) \geq m_{x-1,u}(t_j)$ for all u.

Let x be in the (y,z) place of t_r. Then $y > 1$ (otherwise, for all u, $m_{xu}(t_r) = m_{x-1,u}(t_r) + 1 \geq m_{x-1,u}(t_j) + 1 \geq m_{xu}(t_j)$, contradicting the definition of x). Since $\bar{t}_r$ has $1,2,\ldots,n-1$ in order down successive columns, $x-1$ is in the $(y-1,z)$ place of t_r. Therefore, using Step 1,

$$e_r(x-1,x) = -e_r \quad \text{and} \quad f_r(x-1,x) = -f_r .$$

For $u \geq y$, $m_{xu}(t_r) = m_{x-1,u}(t_r) + 1 \geq m_{x-1,u}(t_j) + 1 \geq m_{xu}(t_j)$.

The definition of x therefore shows that

$$m_{xu}(t_r) < m_{xu}(t_j) \text{ for some } u < y.$$

But $m_{x-1,u}(t_r) = uz$ for $u < y$ (since $\bar{t}_r$ has $1,2,\ldots,n-1$ in order down successive columns), and the first row of t_j contains at most z numbers less than or equal to $x-1$ (since $m_{x-1,1}(t_j) \le m_{x-1,1}(t_r) = z$). Because t_j is standard, this means that x must be in the $(1,z+1)$ place of t_j, and $x-1$ is in a column of t_j no later than the zth column.

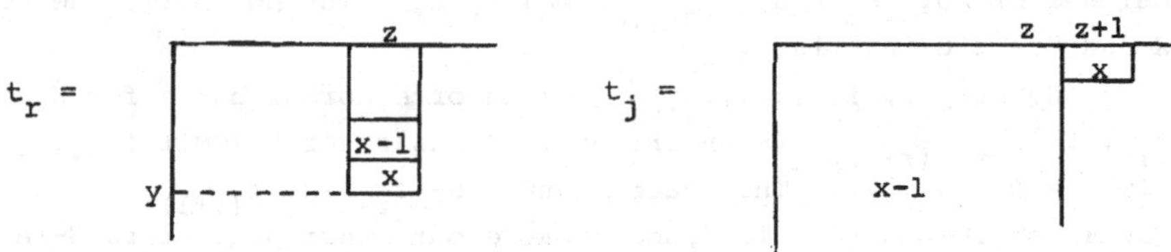

If $t_k = t_j(x-1,x)$, then Step 1 gives

$$f_j(x-1,x) = \sigma_1 f_j + \sigma_2 f_k \quad \text{where } 0 < \sigma_1 < 1.$$

Therefore,

$$b_1 f_1 + \ldots + b_j f_j + \ldots + b_{r-1} f_{r-1} + b_r e_r$$

$$= f_r = -f_r(x-1,x)$$

$$= -b_1 f_1(x-1,x) - \ldots - b_j(\sigma_1 f_j + \sigma_2 f_k) - \ldots + b_r e_r .$$

Since $b_j \neq 0$ and $\sigma_1 \neq -1$, f_j must appear elsewhere in the last line. This means that b_k is non-zero. But $m_{x-1,1}(t_k) = z+1 > z = m_{x-1,1}(t_r)$, and this contradicts our minimal choice of x.

We have thus proved that in the expression

$$f_r = b_1 f_1 + \ldots + b_{r-1} f_{r-1} + b_r e_r$$

$b_j = 0$ unless $\{t_j\} \trianglelefteq \{t_r\}$. Our induction hypothesis at the beginning of Step 2 shows now that f_r is a linear combination of polytabloids e_i with $\{t_i\} \trianglelefteq \{t_r\}$. This concludes the proof of Step 2.

<u>Step 3</u> Calculation of the matrices representing $(n-1,n)$.

Take a new total order on tabloids, containing $\triangleleft$, in which $\{t\}$ and $\{t(n-1,n)\}$ are adjacent if both are standard. (This is possible in view of Lemma 3.16.) We fix our notation by saying that $\{t_1\} < \{t_2\} < \ldots < \{t_d\}$ are the different standard tabloids ordered by definition 3.10, and $\{t_{1\pi}\} \ll \{t_{2\pi}\} \ll \ldots \ll \{t_{d\pi}\}$ is the new order. Thus, π is a permutation of $\{1,2,\ldots,d\}$ and if both $t_{i\pi}$ and $t_{i\pi}(n-1,n)$ are standard then $t_{i\pi}(n-1,n) = t_{(i\pm1)\pi}$.

We plan to evaluate $f_{r\pi}(n-1,n)$. Assume, for the moment, that if $t_{r\pi}(n-1,n)$ is standard, then $t_{r\pi}(n-1,n) = t_{(r+1)\pi}$.

Let G denote the group $\{1,(n-1,n)\}$.

Let X denote the space spanned by $e_{1\pi}, e_{2\pi}, \ldots, e_{(r-1)\pi}$.

Let $Y = X + e_{r\pi}\,\mathbb{R}G$ (so that dim Y = dim X + 2 or 1, depending on whether or not both $t_{r\pi}$ and $t_{r\pi}(n-1,n)$ are standard.)

Since our new total order contains $\trianglelefteq$, for every standard t, neither or both e_t and $e_{t(n-1,n)}$ belong to X (using 25.1). Hence both X and Y are G-invariant.

By Step 2, $f_{1\pi}, \ldots, f_{(r-1)\pi}$ is an orthonormal basis for X and $f_{1\pi}, \ldots, f_{r\pi}, f_{(r+1)\pi}$ is an orthonormal basis for Y (Omit $f_{(r+1)\pi}$ if dim Y = dim X + 1). The space spanned by $f_{r\pi}$ and $f_{(r+1)\pi}$ is the orthogonal complement to X in Y, and because our inner product is G-invariant, the space spanned by $f_{r\pi}$ and $f_{(r+1)\pi}$ is G-invariant (Omit $f_{(r+1)\pi}$ if dim Y = X + 1).

Now, $f_{r\pi}$ = an element of X + b e_r, where b > 0 (since the coefficient of $\{t_{r\pi}\}$ in $f_{r\pi}$ is chosen to be positive). Therefore, when n-1 and n belong to the same row or column of $t_{r\pi}$,

$$f_{r\pi}(n-1,n) = \text{an element of } X + \varepsilon b\, e_{r\pi}$$

$$\text{where } \varepsilon = \begin{cases} +1 \text{ if } n-1 \text{ and } n \text{ are in the same row of } t_{r\pi} \\ -1 \text{ if } n-1 \text{ and } n \text{ are in the same column of } t_{r\pi} \end{cases}$$

But we have just proved that $f_{r\pi}(n-1,n)$ is a multiple of $f_{r\pi}$ in these cases, and comparing coefficients of $e_{r\pi}$, we see that

$$f_{r\pi}(n-1,n) = \varepsilon f_{r\pi}$$

and this completes the case where $t_{r\pi}(n-1,n)$ is not standard.

On the other hand, when both $t_{r\pi}$ and $t_{r\pi}(n-1,n)$ (= $t_{(r+1)\pi}$) are standard,

$$f_{r\pi}(n-1,n) = \text{an element of } X + b\, e_{(r+1)\pi} \qquad (b > 0)$$

Since the space spanned by $f_{r\pi}$ and $f_{(r+1)\pi}$ is G-invariant,

$$f_{r\pi}(n-1,n) = \rho_1 f_{r\pi} + \rho_2 f_{(r+1)\pi}$$

where ρ_1 and ρ_2 are real numbers, and the coefficient of $\{t_{(r+1)\pi}\}$ shows that ρ_2 is strictly positive. Now

$$\langle f_{r\pi}(n-1,n), f_{r\pi}(n-1,n) \rangle = \langle f_{r\pi}, f_{r\pi} \rangle = 1$$

so $\rho_1^2 + \rho_2^2 = 1$ with $\rho_2 > 0$. Also

$$f_{r\pi} = \rho_1 f_{r\pi}(n-1,n) + \rho_2 f_{(r+1)\pi}(n-1,n),$$

whence

$$f_{(r+1)\pi}(n-1,n) = \rho_2 f_{r\pi} - \rho_1 f_{(r+1)\pi}$$

It remains, therefore, to show that ρ_1 may be calculated as in the statement of Young's Orthogonal Form in the case under discussion, where $t_{r\pi}(n-1,n) = t_{(r+1)\pi}$. This will be done using some properties of the group $\mathfrak{S}_3$.

Since n-1 and n are not in the same row or column of $t_{r\pi}$, $n \geq 3$. Also, $t_{r\pi} \lhd t_{r\pi}(n-1,n)$, so n-1 is lower than n in $t_{r\pi}$. There are 4 cases to consider

(i) n-2, n-1 and n appear in $t_{r\pi}$ thus:

n-2	n
n-1	

(ii) Some two numbers from {n-2,n-1,n} are in the same row, but no two are in the same column of $t_{r\pi}$.

(iii) Some two numbers from {n-2,n-1,n} are in the same column, but no two are in the same row of $t_{r\pi}$.

(iv) No two numbers from {n-2,n-1,n} are in the same row or column of $t_{r\pi}$.

We tackle case (ii) first; case (iii) is similar and case (i) is comparatively trivial. Finally we deal with the hard case (iv).

<u>Case (ii)</u> Let H be the group generated by $g_1 = (n-2,n-1)$ and $g_2 = (n-1,n)$. Since n-1 is lower than n in $t_{r\pi}$, $t_{r\pi}$ has the form:

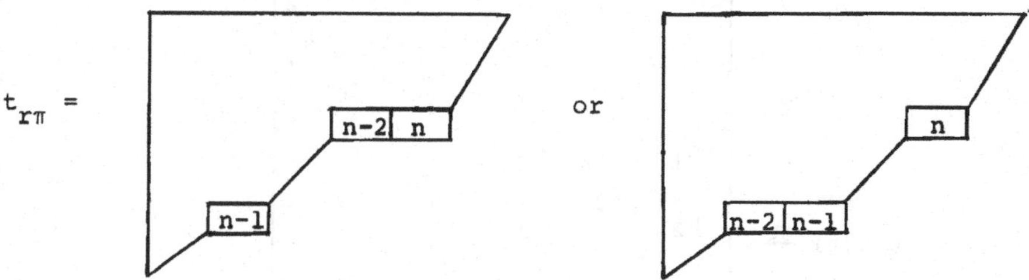

In the first case, let $t = t_{r\pi}$, and in the second let $t = t_{r\pi}(n-1,n)$. The space spanned by f_t, f_{tg_1} and f_{tg_2} is H-invariant. In fact, our results so far show that, with respect to the basis f_t, f_{tg_1}, f_{tg_2} , the action of H on this space is given by

$$g_1 = (n-2,n-1) \leftrightarrow \begin{pmatrix} \sigma_1 & \sigma_2 & 0 \\ \sigma_2 & -\sigma_1 & 0 \\ 0 & 0 & 1 \end{pmatrix} \qquad g_2 = (n-1,n) \leftrightarrow \begin{pmatrix} -\tau_1 & 0 & \tau_2 \\ 0 & 1 & 0 \\ \tau_2 & 0 & \tau_1 \end{pmatrix}$$

where σ_1 is known, from Step 1. The axial distance from n-1 to n in t = -(the axial distance from n-2 to n-1 in t) + 1. We shall therefore have finished if we can prove that $\sigma_1^{-1} = 1 + \tau_1^{-1}$.

Now, trace $g_1 g_2 = -\sigma_1\tau_1 - \sigma_1 + \tau_1$. Therefore

$$|\text{trace } g_1g_2| \le |\sigma_1\tau_1| + |\sigma_1| + |\tau_1| \le \tfrac{1}{2} + \tfrac{1}{2} + 1 = 2.$$

The character table of $\mathfrak{S}_3$ is

	(1^3)	$(2,1)$	(3)
$\chi^{(3)}$	1	1	1
$\chi^{(2,1)}$	2	0	-1
$\chi^{(1^3)}$	1	-1	1

The only representation of dimension 3 having trace 1 on the transpositions and $|\text{trace}| \le 2$ on elements of order 3 is $\chi^{(3)} + \chi^{(2,1)}$. Therefore, trace $g_1g_2 = 0$, giving $\tau_1 = \sigma_1\tau_1 + \sigma_1$, as required.

Case (iv) Let H, g_1 and g_2 be as in Case (ii). We may assume that $n-2$ is higher than $n-1$, and $n-1$ is higher than n in t, and that $t_r = th$ for some h in H. Taking $f_t, f_{tg_1}, f_{tg_2}, f_{tg_2g_1}, f_{tg_2g_1g_2}, f_{tg_2g_1g_2g_1}$ as a basis for $f_t\mathbb{R}H$, g_1 and g_2 are represented by

$$g_1 = (n-2,n-1) \leftrightarrow \begin{pmatrix} -\nu_1 & \nu_2 & & & & \\ \nu_2 & \nu_1 & & & & \\ & & -\omega_1 & \omega_2 & & \\ & & \omega_2 & \omega_1 & & \\ & & & & \pi_1 & \pi_2 \\ & & & & \pi_2 & -\pi_1 \end{pmatrix}$$

$$g_2 = (n-1,n) \leftrightarrow \begin{pmatrix} -\alpha_1 & & \alpha_2 & & & \\ & -\beta_1 & & & & \beta_2 \\ \alpha_2 & & \alpha_1 & & & \\ & & & -\gamma_1 & \gamma_2 & \\ & & & \gamma_2 & \gamma_1 & \\ & \beta_2 & & & & \beta_1 \end{pmatrix}$$

(Omitted entries are zero).

Here we know that each of $\nu_1, \omega_1, \pi_1, \alpha_2, \beta_2, \gamma_2$ is non-zero. The values of ν_1, ω_1 and π_1 are known and $\nu_1^{-1} + \pi_1^{-1} = \omega_1^{-1}$, from Step 1. We want $\alpha_1 = \pi_1$, $\beta_1 = \omega_1$ and $\gamma_1 = \nu_1$. There seems to be no more efficient way of proving this than equating $(g_1g_2)^2$ with g_2g_1, using the fact that g_1g_2 has order 3 (cf. Thrall [23]). The (4,1), (5,2) and (3,1) entries in the relevant matrices give

$$\omega_2\,\alpha_2\,\alpha_1\,\nu_1 - \omega_1\,\omega_2\,\alpha_1\,\alpha_2 - \omega_1\,\omega_2\,\gamma_1\,\alpha_2 = 0$$

$$-\pi_2\,\beta_2\,\nu_1\,\beta_1 + \pi_2\,\pi_1\,\beta_2\,\gamma_1 - \pi_1\,\pi_2\,\beta_1\,\beta_2 = 0$$

and $-\omega_1 \nu_1 \alpha_1 \alpha_2 + \omega_1^2 \alpha_1 \alpha_2 - \omega_2^2 \gamma_1 \alpha_2 = -\alpha_2 \nu_1$

Substituting $\omega_2^2 = 1 - \omega_1^2$ and $\omega_1^{-1} = \nu_1^{-1} + \pi_1^{-1}$, these rapidly give the required result: $\alpha_1 = \pi_1$, $\beta_1 = \omega_1$ and $\gamma_1 = \nu_1$.

This finishes Step 3 and completes the proof of Young's Orthogonal Form.

25.6 EXAMPLE Here is the orthonormal basis of $S_{\mathbb{R}}^{(3,2)}$ in terms of the graphs used in Example 5.2:

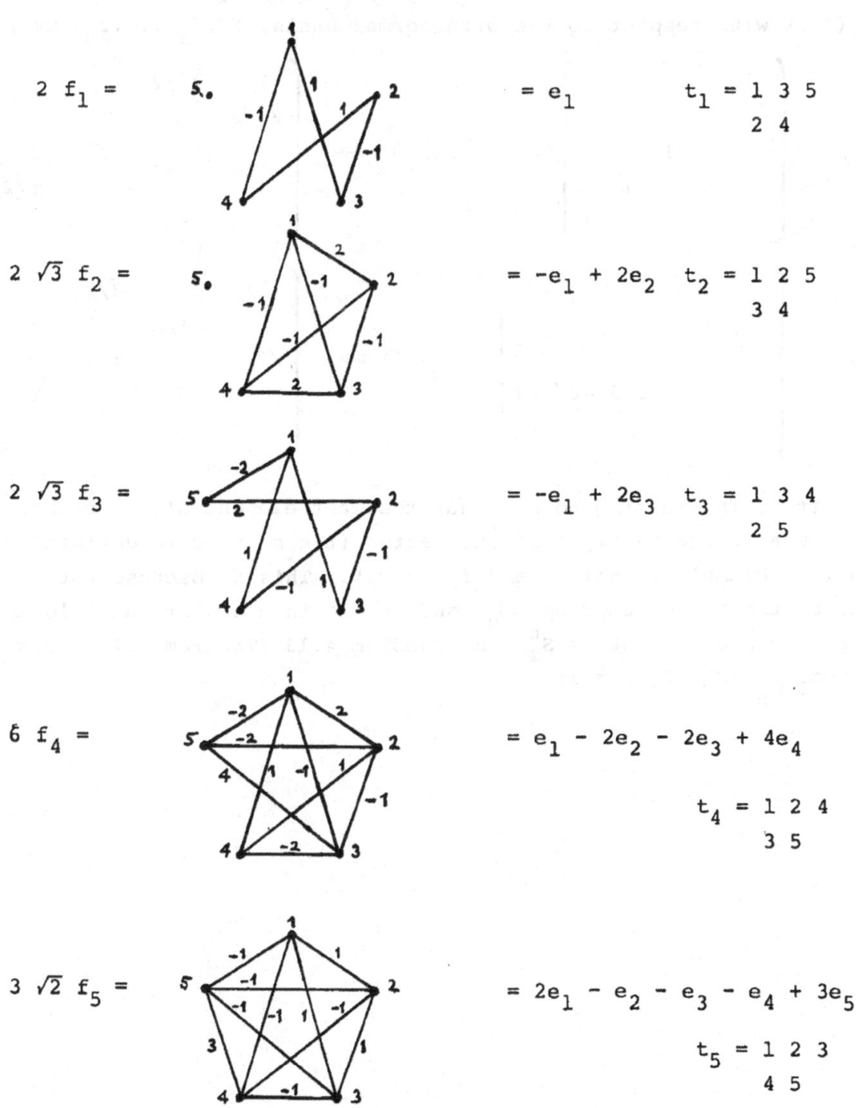

For clarity, we have chosen the graphs (= $G_1, G_2, \ldots, G_5$, say) so that the edges have integer coefficients. It is easy to check that the graphs are orthogonal, and that $\{t_i\}$ is the last tabloid involved in G_i. The numbers multiplying each f_i ensure that $< f_i, f_i > = 1$ (For example, $< G_3, G_3 > = 1\ 2$, so $(2\sqrt{3})^{-1}G_3$ has norm 1).

Corollary 8.12 has been used to write the graphs in terms of polytabloids. Since $\{t_2\} \not\trianglerighteq \{t_3\}$, e_2 is not involved in f_3, illustrating Theorem 25.3.

Writing out in full the matrices representing (1 2),(2 3),(3 4) and (4 5) with respect to the orthonormal basis, $f_1, f_2, \ldots, f_5$, we have:

$$(1\ 2) \leftrightarrow \begin{pmatrix} -1 & & & & \\ & 1 & & & \\ & & -1 & & \\ & & & 1 & \\ & & & & 1 \end{pmatrix} \qquad (2\ 3) \leftrightarrow \begin{pmatrix} 1/2 & \sqrt{3}/2 & & & \\ \sqrt{3}/2 & -1/2 & & & \\ & & 1/2 & \sqrt{3}/2 & \\ & & \sqrt{3}/2 & -1/2 & \\ & & & & 1 \end{pmatrix}$$

$$(3\ 4) \leftrightarrow \begin{pmatrix} -1 & & & & \\ & 1 & & & \\ & & 1 & & \\ & & & 1/3 & 2\sqrt{2}/3 \\ & & & 2\sqrt{2}/3 & -1/3 \end{pmatrix} \qquad (4\ 5) \leftrightarrow \begin{pmatrix} 1/2 & & \sqrt{3}/2 & & \\ & 1/2 & & \sqrt{3}/2 & \\ \sqrt{3}/2 & & -1/2 & & \\ & \sqrt{3}/2 & & -1/2 & \\ & & & & 1 \end{pmatrix}$$

It is interesting to see that the last element of the orthonormal basis is always a multiple of the vector $\{t\}\kappa_t\rho_t$ used in definition 23.3 (cf. Example 23.6(iii) and f_5 above). This is because both are fixed by the Young subgroup $\mathfrak{S}_\mu$ and to within a scalar multiple $\mathfrak{S}_\mu$ fixes a unique element of $S^\mu_{\mathbb{R}}$, by Theorem 4.13 (Theorem 4.13 shows that $\dim \mathrm{Hom}_{\mathbb{R}\mathfrak{S}_n}(M^\mu_{\mathbb{R}}, S^\mu_{\mathbb{R}}) = 1$).

26 REPRESENTATIONS OF THE GENERAL LINEAR GROUP

The representation theory of $\mathfrak{S}_n$ is useful in the study of more general permutation groups. For example, Frobenius used part of the character table of $\mathfrak{S}_{24}$ to find that of the Mathieu group M_{24}. There is another, less obvious application of the theory, following from a study of the group $GL_d(F)$ of non-singular $d \times d$ matrices over a field F. Remember that any group which has a representation of dimension d over F has (by definition) a homomorphic image inside $GL_d(F)$. Although the results of this section will be stated in terms of the general linear group, they apply equally well to any subgroup thereof. We plan to construct, for each n and each partition of n, a representation of $GL_d(F)$ over F. Hence from any representation of any group, we can produce infinitely many new representations over the same field.

$GL_d(F)$ acts naturally on a d-dimensional vector space, $W^{(1)}$ say, over F. Let $\overline{\underline{1}}, \overline{\underline{2}}, \ldots, \overline{\underline{d}}$ be a basis for $W^{(1)}$. If $g = (g_{ij})$ is a matrix in $GL_d(F)$, then

$$\overline{\underline{i}}\, g = \sum_j g_{ij}\, \overline{\underline{j}} \,.$$

The general element of $W^{(1)} \otimes W^{(1)}$ may be written as

$$\sum_{i,j \le d} a_{ij} \begin{matrix}\overline{\underline{i}}\\ \underline{j}\end{matrix} \quad (a_{ij} \in F) \,.$$

(The reason for this perverse notation will emerge later.) Let $GL_d(F)$ act on $W^{(1)} \otimes W^{(1)}$ by

$$\sum_{i,j \le d} a_{ij} \begin{matrix}\overline{\underline{i}}\\ \underline{j}\end{matrix} g = \sum_{i,j,k,\ell} a_{ij} g_{ik} g_{j\ell} \begin{matrix}\overline{\underline{k}}\\ \underline{\ell}\end{matrix} \qquad (g \in GL_d(F)),$$

as usual.

For the moment, assume <u>char F = 0</u>. There are two natural $GL_d(F)$-invariant subspaces of $W^{(1)} \otimes W^{(1)}$, namely those spanned by

$$\left\{ \begin{matrix}\overline{\underline{i}}\\ \underline{j}\end{matrix} + \begin{matrix}\overline{\underline{j}}\\ \underline{i}\end{matrix} \;\middle|\; 1 \le i \le j \le d \right\}$$

and by

$$\left\{ \begin{matrix}\overline{\underline{i}}\\ \underline{j}\end{matrix} - \begin{matrix}\overline{\underline{j}}\\ \underline{i}\end{matrix} \;\middle|\; 1 \le i < j \le d \right\} .$$

These are called the symmetric part of $W^{(1)} \otimes W^{(1)}$ and the second exterior power of $W^{(1)}$ (or the skew-symmetric part of $W^{(1)} \otimes W^{(1)}$) respectively. Since char F = 0

$$W^{(1)} \otimes W^{(1)} = (\text{symmetric part}) \oplus (\text{2nd exterior power}).$$

Write this as

$$W^{(1)} \otimes W^{(1)} \cong W^{(2)} \oplus W^{(1^2)} .$$

Less wellknown is that

$$W^{(1)} \otimes W^{(1)} \cong W^{(3)} \oplus 2W^{(2,1)} \oplus W^{(1^3)}$$

for some subspaces $W^{(3)}$ (called the 3rd symmetric power), $W^{(2,1)}$ (of which there are two copies) and $W^{(1^3)}$ (called the 3rd exterior power)

Also
$$W^{(1)}\otimes W^{(1)}\otimes W^{(1)} \otimes W^{(1)} \cong W^{(4)}\oplus 3W^{(3,1)}\oplus 2W^{(2,2)}\oplus 3W^{(2,1^2)} \oplus W^{(1^4)}$$

"and so on". Further
$$W^{(2)}\otimes W^{(2)}\cong W^{(4)} \oplus W^{(3,1)}\oplus W^{(2,2)}.$$

Most of the work needed to prove these results has already been done, since they are similar to those for the symmetric group (compare the last example with $S^{(2)}\otimes S^{(2)}\uparrow \mathfrak{S}_4 \cong S^{(4)} \oplus S^{(3,1)}\oplus S^{(2,2)}$, when char F = 0).

Consider again $W^{(1)}\otimes W^{(1)}$. How do we deal with the symmetric and skew-symmetric parts when F is arbitrary (allowing char F = 2)? We adjust our notation, by letting $W^{(2)}$ be the space of homogeneous polynomials of degree 2 in <u>commuting</u> variables $\underline{1}, \underline{2},\ldots,\underline{d}$. We write

$$\overline{\underline{i\ j}} \text{ for the monomial } \underline{i}\ \underline{j}$$

so that
$$\overline{\underline{i\ j}} = \overline{\underline{j\ i}} \text{ and } W^{(2)} \text{ is spanned by } \{\overline{\underline{i\ j}}\ |1 \le i \le j \le d\}.$$

We keep our previous notation for $W^{(1)} \otimes W^{(1)}$ and for $W^{(1^2)}$, and now
$$(W^{(1)}\otimes W^{(1)})/W^{(1^2)} \cong W^{(2)} \text{ as vector spaces, since}$$
$$\frac{\overline{\underline{i}}}{\underline{j}} \equiv \frac{\overline{\underline{j}}}{\underline{i}} \text{ modulo } W^{(1^2)}.$$

Another way of looking at this is to define the linear transformation $\psi_{1,0} : W^{(1)}\otimes W^{(1)}\to W^{(2)}$ by
$$\frac{\overline{\underline{i}}}{\underline{j}} \to \overline{\underline{i\ j}}\ .$$

Then $\ker \psi_{1,0} = W^{(1^2)}$. If we let $GL_d(F)$ act on $W^{(2)}$ in the natural way then $\psi_{1,0}$ turns out to be a $GL_d(F)$-homomorphism:

$$\frac{\overline{\underline{i}}}{\underline{j}}\, g = \sum_{k,\ell} g_{ik}g_{j\ell} \frac{\overline{\underline{k}}}{\underline{\ell}} \underset{\psi_{1,0}}{\longrightarrow} \sum_{k,\ell} g_{ik}g_{j\ell}\ \overline{\underline{k\ \ell}} = \overline{\underline{i\ j}}\, g\ .$$

It is the generalization of $W^{(2)}$, described in the way above, to the kth symmetric power of $W^{(1)}$ which we take as our building block for the representation theory of $GL_d(F)$.

26.1 DEFINITION The <u>kth symmetric power</u> of $W^{(1)}$ is the vector space $W^{(k)}$ of homogeneous polynomials of degree k in commuting variables $\underline{1}, \underline{2},\ldots,\underline{d}$, with coefficients from F. We write
$$\overline{\underline{i_1\ i_2 \ldots i_k}} \text{ for the monomial } \underline{i}_1\ \underline{i}_2 \cdots \underline{i}_k$$

and we let the $G\,L_d(F)$ action on $W^{(k)}$ be defined by

$$\overline{i_1 i_2 \ldots i_k}\; g = \Sigma\, g_{i_1 j_1} g_{i_2 j_2} \ldots g_{i_k j_k}\; \overline{j_1 j_2 \ldots j_k}$$

where the sum is over all suffices $j_1, j_2, \ldots, j_k$ between 1 and d, and $g = (g_{ij})$.

The reader who is more familiar with the kth symmetric power as the subspace $\mathrm{Symm}_k(W^{(1)})$ of $W^{(1)} \otimes \ldots \otimes W^{(1)}$ (k times) spanned by certain symmetrized vectors, may find it useful to know that the connection between this and $W^{(k)}$ is:

$$W^{(k)*} = \mathrm{Symm}_k(W^{(1)*}),$$

where * denotes the process of taking duals.

Corresponding to $M^\mu = S^{0,\mu}$ in the representation theory of $\mathfrak{S}_n$, we consider the space $W^{(\mu_1)} \otimes \ldots \otimes W^{(\mu_n)}$. There is still a little more preliminary work, though, before we come to this. It should, however, be clear that it is useful to discuss vector spaces spanned by tabloids with repeated entries (For the time being, it is best to forget any intended interpretation in terms of the action of $G\,L_d(F)$).

Let $X = x_1\, x_2 \ldots x_n$ be a sequence of non-decreasing positive integers. If μ is a partition of n and t is a μ-tableau (of type (1^n)) let $t\,\bar{X}$ denote the array of integers obtained by making the substitutions $i \to x_i$ in t $(1 \le i \le n)$. Let $t_1\,\bar{X} \sim t_2\bar{X}$ if and only if for all m and r the number of m's in the rth row of $t_1\bar{X}$ equals the number of m's in the rth row of $t_2\bar{X}$, and let $\{t\bar{X}\}$ denote the $\sim$-class containing $t\bar{X}$. Then

$$\{t\} \to \{t\}\bar{X} = \{t\bar{X}\}$$

is clearly a well-defined map from the set of μ-tabloids of type (1^n) onto the set of μ-tabloids of type ξ, where the partition ξ is defined by

$$\xi_i = \text{the number of terms of } X \text{ equal to } i.$$

(As in some of our earlier work, we do not require μ and ξ to be proper partitions of n.) Extend $\bar{X}$ to be a linear map on $S^{0,\mu}$, the space spanned by the μ-tabloids.

26.2 EXAMPLES (i) If X = 1 1 2, then

$S^{0,(2,1)}\bar{X}$ is spanned by $\begin{array}{l}\overline{1\;1}\\ \underline{2}\end{array}$ and $\begin{array}{l}\overline{1\;2}\\ \underline{1}\end{array}$

$S^{(2,1),(2,1)}\bar{X}$ is spanned by $\begin{array}{l}\overline{1\;1}\\ \underline{2}\end{array} - \begin{array}{l}\overline{2\;1}\\ \underline{1}\end{array}$

(ii) If X = 1 1 1, then

$S^{0,(2,1)}\bar{X}$ is spanned by $\begin{array}{l}\overline{1\;1}\\ \underline{1}\end{array}$

$S^{(2,1),(2,1)}\bar{X} = 0.$

Certain linear transformations $\psi_{i,v}$ were defined on the vector spaces $S^{O,\mu}$ in section 17. Define the corresponding linear transformations on $S^{O,\mu}\bar{X}$ by

$$\{t\}\bar{X}\,\psi_{i,v} = \{t\}\psi_{i,v}\,\bar{X}\ .$$

(It is clear that this is welldefined.)

26.3 THEOREM <u>Suppose that X is a sequence of type ξ, λ is a proper partition, and $\mu^{\#},\mu$ are a pair of partitions as in 15.5.</u>
Then

(i) <u>$\dim S^{\lambda}\bar{X}$ = the number of semistandard λ-tableaux of type ξ</u>

(ii) <u>$S^{\mu^{\#},\mu}\bar{X}\,\psi_{c-1,\mu_c^{\#}} = S^{\mu^{\#},\mu R_c}\,\bar{X}$</u>

(iii) <u>$S^{\mu^{\#},\mu}\,\bar{X} \cap \ker\psi_{c-1,\mu_c^{\#}} = S^{\mu^{\#}A_c,\mu}\,\bar{X}$</u> .

<u>Proof</u>: In 17.12, we proved that

$$e_t^{\mu^{\#},\mu}\,\psi_{c-1,\mu_c^{\#}} = e_{tR_c}^{\mu^{\#},\mu R_c} \text{ and } e_t^{\mu^{\#}A_c,\mu}\,\psi_{c-1,\mu_c^{\#}} = 0\ .$$

Applying $\bar{X}$ to these equations, we deduce that

$$S^{\mu^{\#},\mu}\,\bar{X}\,\psi_{c-1,\mu_c^{\#}} = S^{\mu^{\#},\mu R_c}\,\bar{X}$$

and $$S^{\mu^{\#}A_c,\mu}\,\bar{X}\,\psi_{c-1,\mu_c^{\#}} = 0.$$

By considering last tabloids, as in the construction of the standard basis of the Specht module, obviously $\dim S^{\lambda}\,\bar{X} \geq |\mathcal{T}_o(\lambda,\xi)|$, where $\mathcal{T}_o(\lambda,\xi)$ is the set of semistandard λ-tableaux of type ξ. If this inequality is strict for some λ, or if $S^{\mu^{\#},\mu}\,\bar{X} \cap \ker\psi_{c-1,\mu_c^{\#}}$ strictly contains $S^{\mu^{\#}A_c,\mu}\bar{X}$ for some pair of partitions $\mu^{\#},\mu$, then choos a pair of partitions $0,\nu$ and a sequence of operations A_c,R_c leading from $0,\nu$ to λ,λ or $\mu^{\#},\mu$, respectively (using 15.12). For each proper partition σ of n, let a_σ be the multiplicity of $S_{\mathbb{C}}^{\sigma}$ as a factor of $S_{\mathbb{C}}^{O,\nu}$. Then there is a series of subspaces of $S^{O,\nu}\,\bar{X}$ with at least a_σ factors isomorphic to $S^{\sigma}\bar{X}$ (cf. Corollary 17.14).
Therefore,

$$\begin{aligned}\text{the number of } \nu\text{-tabloids of type } \xi &= \dim S^{O,\nu}\,\bar{X}\\ &\geq \sum_{\sigma} a_\sigma \dim S^{\sigma}\,\bar{X}\\ &\geq \sum_{\sigma} a_\sigma\,|\mathcal{T}_o(\sigma,\xi)|\\ &= \sum_{\sigma} a_\sigma \dim \operatorname{Hom}_{\mathbb{C}\mathfrak{S}_n}(S_{\mathbb{C}}^{\sigma},M_{\mathbb{C}}^{\xi}),\ \text{by Corollary 13.14.}\end{aligned}$$

At least one of the inequalities is strict (the first is strict i our kernel is too big, and the second is strict if $\dim S^{\lambda} > |\mathcal{T}_o(\lambda,\xi)|$
Recall that a_σ is the multiplicity of $S_{\mathbb{C}}^{\sigma}$ as a factor of $M_{\mathbb{C}}^{\nu} = S_{\mathbb{C}}^{O,\nu^{O}}$.

Therefore,

$$\sum_{\sigma} a_\sigma \dim \operatorname{Hom}_{\mathbb{C}\mathfrak{S}_n}(S^\sigma_{\mathbb{C}}, M^\xi_{\mathbb{C}}) = \dim \operatorname{Hom}_{\mathbb{C}\mathfrak{S}_n}(M^\nu_{\mathbb{C}}, M^\xi_{\mathbb{C}})$$

= the number of ν-tabloids of type ξ, by Theorem 13.19.

This contradiction completes the proof.

26.4 DEFINITIONS Let $W^{\mu^\#,\mu}$ be the vector space direct sum of $S^{\mu^\#,\mu}\,\bar{X}$ where $\bar{X}$ runs over all non-decreasing sequences whose terms are $1,2,\ldots,d$. Let the ψ maps act on $W^{\mu^\#,\mu}$ by acting on each component separately. When μ is a proper partition of n, let $W^\mu = W^{\mu,\mu}$.

We now have

26.5 THEOREM <u>Let λ be a proper partition of n. Then</u>

<u>(i) dim W^λ equals the number of semistandard λ-tableaux with entries from $\{1,2,\ldots,d\}$</u>

<u>(ii) W^λ is an intersection of kernels of ψ-maps defined on $W^{0,\lambda}$</u>.

<u>Proof</u>: This follows immediately from Theorem 26.3, since W^λ is the direct sum of the spaces $S^\lambda\,\bar{X}$.

Next, identify $W^{0,\mu}$ with $W^{(\mu_1)} \otimes W^{(\mu_2)} \otimes \ldots \otimes W^{(\mu_n)}$. We have defined the action of $G\,L_d(F)$ on a symmetric power, and hence $G\,L_d(F)$ acts on $W^{0,\mu}$. An unpleasant use of suffix notation shows that the ψ-maps commute with the action of $G\,L_d(F)$, and then Theorem 26.5 shows that W^λ is a $G\,L_d(F)$ module, which we call a <u>Weyl module</u>.

From Theorem 26.3, we have

26.6 THEOREM <u>$W^{\mu^\#,\mu}$ has a series, all of whose factors are Weyl modules. The number of times W^λ occurs in this series equals the number of times the Specht module S^λ occurs in a Specht series for $S^{\mu^\#,\mu}$</u>.

In particular, the number of times W^λ occurs in a Weyl series for $W^{0,\mu} = W^{(\mu_1)} \otimes W^{(\mu_2)} \otimes \ldots \otimes W^{(\mu_n)}$ is given by Young's Rule. (Notice that no "inducing up" takes place here, as it did in the corresponding symmetric group case). This justifies all the examples we gave at the beginning of the section; indeed, we have proved their characteristic-free analogues. For example, $W^{(1)} \otimes W^{(1)} \otimes W^{(1)}$ has a $G\,L_d(F)$ series with factors isomorphic to $W^{(3)}, W^{(2,1)}, W^{(2,1)}, W^{(1^3)}$, in order from the top, and this holds for every field F.

We now investigate character values. Let

$$g = \begin{pmatrix} \alpha_1 & & 0 \\ & \alpha_2 & \\ * & & \ddots \\ & & & \alpha_d \end{pmatrix} \in G\,L_d(F)$$

If F is algebraically closed, every elements of $GL_d(F)$ is conjugate to one of the above form, and so it is sufficient to specify the character of g on a Weyl module.

26.7 DEFINITION For an integer k, let {k} denote the kth homogeneous symmetric function of $\alpha_1,\ldots,\alpha_d$. That is,

$$\{k\} = \sum_{1\le i_1\le \ldots \le i_k \le d} \alpha_{i_1}\alpha_{i_2}\ldots\alpha_{i_k}$$

(By convention {0} = 1 and {k} = 0 if k < 0)

26.8 EXAMPLES $\{1\} = \alpha_1 + \alpha_2 + \ldots + \alpha_d$

$$\{2\} = \alpha_1^2 + \alpha_2^2 + \ldots + \alpha_d^2 + \alpha_1\alpha_2 + \alpha_1\alpha_3 + \ldots + \alpha_{d-1}\alpha_d$$

$$\{3\} = \alpha_1^3 + \ldots + \alpha_d^3 + \alpha_1^2\alpha_2 + \alpha_1\alpha_2^2 + \ldots + \alpha_{d-1}^2\alpha_d + \alpha_{d-1}\alpha_d^2 + \alpha_1\alpha_2\alpha_3 + \ldots + \alpha_{d-2}\alpha_{d-1}\alpha_d$$

26.9 THEOREM <u>{k} is the character of g on $W^{(k)}$</u>.

<u>Proof</u> $\bar{\underline{i}}\, g = \alpha_i \bar{\underline{i}}$ + a combination of $\bar{\underline{j}}$'s with j < i. Therefore, if $1 \le i_1 \le \ldots \le i_k \le d$, then the coefficient of $\overline{\underline{i_1\ldots i_k}}$ in $\overline{\underline{i_1\ldots i_k}}\, g$ is $\alpha_{i_1}\ldots\alpha_{i_k}$. Since $W^{(k)}$ has a basis consisting of elements of the form $\overline{\underline{i_1\ldots i_k}}$, the result follows.

26.10 COROLLARY <u>$\{\lambda_1\}\ldots\{\lambda_n\}$ is the character of g on $W^{(\lambda_1)} \otimes \ldots \otimes W^{(\lambda_n)} = W^{0,\lambda}$</u>

Now, recall from 6.1 that $m = (m_{\lambda\mu})$ is the matrix whose entries are indexed by proper partitions, given by

$$[\lambda_1][\lambda_2]\ldots[\lambda_n] = \sum_\mu m_{\lambda\mu}[\mu] .$$

From Theorem 26.6, we have

<u>26.11</u> $$\{\lambda_1\}\{\lambda_2\}\ldots\{\lambda_n\} = \sum_\mu m_{\lambda\mu}\{\mu\}.$$

Since the Determinantal Form gives the inverse of the matrix m, we have

26.12 THEOREM <u>If λ is a proper partition of n, then the character of g on the Weyl module W^λ is $|\{\lambda_i - i + j\}|$</u>.

We write $\{\lambda\} = |\{\lambda_i - i + j\}|$ = the character of g on W^λ. Then immediately

26.13 THEOREM <u>$\{\lambda\}\{\mu\}$ is the character of g on $W^\lambda \otimes W^\mu$</u>.

The Littlewood-Richardson Rule tells us how to evaluate $\{\lambda\}\{\mu\}$ as a linear combination of $\{\nu\}$'s (where λ is a partition of r, μ is a is a partition of n-r and ν is a partition of n), since we know that the Littlewood-Richardson Rule follows from Young's Rule.

It is worth noting that were we to define

$$\{k\} = \sum_{1\le i_1 \le \ldots \le i_k} \alpha_{i_1} \alpha_{i_2} \ldots \alpha_{i_k}$$

where $\{\alpha_1, \alpha_2, \ldots\}$ is countable set of indeterminates, then

$$\{\lambda_1\}\{\lambda_2\}\ldots\{\lambda_n\} = \sum_{\lambda} m_{\lambda\mu}\{\mu\}$$

and $\{\lambda\} = |\{\lambda_i - i + j\}|$

are equivalent definitions of $\{\lambda\}$, for λ a partition of n (since our results work for $\alpha_1, \ldots, \alpha_d$ in an infinite field, the above must be identities in the indeterminates $\alpha_1, \ldots, \alpha_d$).

$\{\lambda\}$ is called a <u>Schur function</u>, and the algebra of Schur functions is thus isomorphic to the algebra generated by the $[\lambda]$'s, where λ varies over partitions of various n. The Littlewood-Richardson Rule enables us to multiply Schur functions.

Schur functions can be evaluated explicitly by

26.14 THEOREM <u>If μ is a proper partition of n, then</u>

$$\{\mu\} = \sum_{\nu} m_{\nu\mu} \, \Sigma' \, \alpha_{i_1}^{\nu_1} \alpha_{i_2}^{\nu_2} \ldots \alpha_{i_n}^{\nu_n}$$

<u>Note</u>: In all that follows, Σ' denotes the sum over all unordered sets of n indices $i_1, i_2, \ldots, i_n$ (no two equal) chosen from $\{1,2,\ldots,d\}$ or from $\{1,2,\ldots\}$ depending on whether we wish to define $\{\mu\}$in terms of $\{\alpha_1, \alpha_2, \ldots, \alpha_d\}$ or of $\{\alpha_1, \alpha_2, \ldots\}$.

<u>Proof of Theorem 26.14</u> $(m\, m')_{\lambda\nu} = \sum_{\sigma} m_{\lambda\sigma} m_{\nu\sigma}$

$= (\sum_{\sigma} m_{\lambda\sigma} \chi^{\sigma}, \sum_{\tau} m_{\nu\tau} \chi^{\tau})$, this being an inner product of characters of $\mathfrak{S}_n$.

$= (\chi^{[\lambda_1][\lambda_2]\ldots[\lambda_n]}, \chi^{[\nu_1][\nu_2]\ldots[\nu_n]})$, by the definition of m.

$= \dim \operatorname{Hom}_{\mathbb{C}\mathfrak{S}_n}(M^{\lambda}, M^{\nu})$

= the number of λ-tabloids of type ν, by Theorem 13.19.

= the coefficient of $\alpha_1^{\nu_1}\alpha_2^{\nu_2}\ldots\alpha_n^{\nu_n}$ in $\{\lambda_1\}\ldots\{\lambda_n\}$, by considering how this coefficient is evaluated,

Therefore, $\{\lambda_1\}\ldots\{\lambda_n\} = \sum_{\nu} (m\, m')_{\lambda\nu} \, \Sigma' \, \alpha_{i_1}^{\nu_1} \alpha_{i_2}^{\nu_2} \ldots \alpha_{i_n}^{\nu_n}$.

But $\{\mu\} = \sum_{\lambda} (m^{-1})_{\mu\lambda} \{\lambda_1\}\ldots\{\lambda_n\}$ by 26.11,

$$= \sum_{\lambda,\nu,\sigma} (m^{-1})_{\mu\lambda} m_{\lambda\sigma} m_{\nu\sigma} \, \Sigma' \, \alpha_{i_1}^{\nu_1} \alpha_{i_2}^{\nu_2} \ldots \alpha_{i_n}^{\nu_n}$$

$$= \sum_{\nu} m_{\nu\mu} \Sigma' \alpha_{i_1}^{\nu_1} \alpha_{i_2}^{\nu_2} \cdots \alpha_{i_n}^{\nu_n}$$, as required.

26.15 DEFINITION Let $s_k = \sum_i \alpha_i^k$ if $k \geq 1$ and $s_0 = 1$.

We can now prove the useful

26.16 THEOREM Let ρ be a permutation of $\mathfrak{S}_n$ with cycle lengths $\rho_1, \rho_2, \ldots, \rho_n$ and let $C(\rho)$ denote the centraliser of ρ in $\mathfrak{S}_n$. Let $\chi^\mu(\rho)$ be the value of the character of $\mathfrak{S}_n$ corresponding to the partition μ, evaluated on ρ. Then

$$\text{(i)} \quad s_{\rho_1} s_{\rho_2} \cdots s_{\rho_n} = \sum_{\mu} \chi^\mu(\rho)\{\mu\}$$

$$\text{(ii)} \quad \{\mu\} = \sum_{\rho} \frac{1}{|C(\rho)|} \chi^\mu(\rho) s_{\rho_1} s_{\rho_2} \cdots s_{\rho_n} .$$

Proof $\chi^{[\nu_1][\nu_2]\ldots[\nu_n]}(\rho)$ = the number of tabloids in M^μ fixed by ρ.
= the number of ν-tabloids of type (1^n) where each cycle of ρ is contained in a single row of the tabloid.
= the coefficient of $\alpha_1^{\nu_1} \alpha_2^{\nu_2} \cdots \alpha_n^{\nu_n}$ in $s_{\rho_1} s_{\rho_2} \cdots s_{\rho_n}$, by considering how this coefficient is evaluated.

Therefore, $s_{\rho_1} s_{\rho_2} \cdots s_{\rho_n} = \sum_{\nu} \chi^{[\nu_1][\nu_2]\ldots[\nu_n]}(\rho) \Sigma' \alpha_{i_1}^{\nu_1} \alpha_{i_2}^{\nu_2} \cdots \alpha_{i_n}^{\nu_n}$

$$= \sum_{\nu,\mu} \chi^{[\nu_1][\nu_2]\ldots[\nu_n]}(\rho)(m_{\mu\nu})^{-1}\{\mu\}, \text{ by Theorem 26.14}$$

$$= \sum_{\mu} \chi^\mu(\rho) \{\mu\}, \text{ from the definition of } m.$$

This proves part (i) of the Theorem.

By the orthogonality relations for the columns of the character table of $\mathfrak{S}_n$,

$$\sum_{\rho} \frac{1}{|C(\rho)|} \chi^\lambda(\rho) s_{\rho_1} s_{\rho_2} \cdots s_{\rho_n} = \sum_{\mu,\rho} \frac{1}{|C(\rho)|} \chi^\lambda(\rho)\chi^\mu(\rho)\{\mu\} = \{\lambda\},$$

and this is the second part of the Theorem.

26.17 COROLLARY If G is any group, and Θ is an ordinary character of G, then for all $n \geq 0$ and all proper partitions μ of n, Θ^μ is a character of G, where

$$\Theta^\mu(g) = \sum_{\rho} \frac{1}{|C(\rho)|} \chi^\mu(\rho) \; \Theta(g^{\rho_1})\Theta(g^{\rho_2})\ldots\Theta(g^{\rho_u}) \quad (g \in G).$$

The centraliser order $|C(\rho)|$ and the character χ^μ refer to the symmetric group $\mathfrak{S}_n$ and the sum is over all proper partitions ρ of n; ρ is written as $(\rho_1, \rho_2, \ldots, \rho_u)$, where $\rho_1 \geq \rho_2 \geq \ldots \geq \rho_u > 0$.

If Θ has degree d, then Θ^μ has degree equal to the number of semistandard μ-tableaux with entries from $\{1,2,\ldots,d\}$.

Proof: There is a homomorphism ϕ from G into $GL_d(C)$. If $g \in G$, let $\phi(g)$ have eigenvalues $\alpha_1, \alpha_2, \ldots, \alpha_d$. Then $\alpha_1^k, \alpha_2^k, \ldots, \alpha_d^k$ are the eigenvalues of g^k, and so $\Theta(g^k) = \alpha_1^k + \ldots + \alpha_d^k$. The result now follows from Theorem 26.16(ii) and Theorem 26.5(i).

26.18 EXAMPLES Referring to the character tables of $\mathfrak{S}_0, \mathfrak{S}_1, \mathfrak{S}_2$ and $\mathfrak{S}_3$, the last of which is

	(1^3)	$(2,1)$	(3)
Centraliser order:	6	2	3
$\chi^{(3)}$	1	1	1
$\chi^{(2,1)}$	2	0	-1
$\chi^{(1^3)}$	1	-1	1

we have, for any ordinary character Θ of any group G, and any g in G,

$$\Theta^{(0)} = \text{the trivial character of } G$$
$$\Theta^{(1)} = \Theta$$
$$\Theta^{(2)}(g) = \tfrac{1}{2}(\Theta(g))^2 + \tfrac{1}{2}\Theta(g^2)$$
$$\Theta^{(1^2)}(g) = \tfrac{1}{2}(\Theta(g))^2 - \tfrac{1}{2}\Theta(g^2)$$
$$\Theta^{(3)}(g) = \tfrac{1}{6}(\Theta(g))^3 + \tfrac{1}{2}\Theta(g^2)\Theta(g) + \tfrac{1}{3}\Theta(g^3)$$
$$\Theta^{(2,1)}(g) = \tfrac{1}{3}(\Theta(g))^3 + 0.\Theta(g^2)\Theta(g) - \tfrac{1}{3}\Theta(g^3)$$
$$\Theta^{(1^3)}(g) = \tfrac{1}{6}(\Theta(g))^3 - \tfrac{1}{2}\Theta(g^2)\Theta(g) + \tfrac{1}{3}\Theta(g^3).$$

Note that $\Theta^{(1)} \otimes \Theta^{(1)} = \Theta^{(2)} + \Theta^{(1^2)}$

$\Theta^{(2)} \otimes \Theta^{(1)} = \Theta^{(2,1)} + \Theta^{(3)}$, etc. (cf. Young's Rule)

If Θ has degree d, then

$$\deg \Theta^{(2)} = \binom{d}{2} + d = \frac{d(d+1)}{2}$$
$$\deg \Theta^{(1^2)} = \binom{d}{2} = \frac{d(d-1)}{2}$$
$$\deg \Theta^{(1^3)} = \binom{d}{3}$$
$$\deg \Theta^{(2,1)} = \frac{(d+1)d(d-1)}{3}$$
$$\deg \Theta^{(3)} = \binom{d+2}{3}$$

(The last two degrees are **most easily** calculated by using the next Theorem.)

Similar to the Hook Formula for dim S^λ, we have

26.19 THEOREM $\dim W^\lambda = \dfrac{\prod_{(i,j)\in[\lambda]} (d+j-1)}{\prod(\text{hook lengths in } [\lambda])}$.

Proof: We prove first that $\dim W^{(k)} = \binom{k+d-1}{d-1}$ if k is a non-negative integer.

The natural basis of $W^{(k)}$ consists of (k)-tabloids with entries from {1,2,...,d}. There is a 1-1 correspondence between this basis and sequences of "bars" (|) and "stars" (*) with d-1 bars and k stars

e.g. * | | * * | * | | | * * * | *
↔ 1 3 3 4 7 7 7 8

There are $\binom{k+d-1}{d-1}$ such sequences, so this is the dimension of $W^{(k}$

Since $\{\lambda\} = |\{\lambda_i + j - i\}|$, we have

$$\dim W^{\lambda} = \left| \binom{\lambda_i + d - 1 + j - i}{d-1} \right| = \left| \binom{\lambda_i + d - 1 + j - i}{\lambda_i + j - i} \right|$$

$$= \left| \frac{d(d+1)\dots(d+\lambda_i - 1 + j - i)}{(\lambda_i + j - i)!} \right| = f(d), \text{ say.}$$

Let λ have h non-zero parts (so we are taking the determinant of an $h \times h$ matrix). It is clear that the polynomial f(d) has degree $\lambda_1 + \lambda_2 + \dots + \lambda_h$ and leading coefficient

$$\left| \frac{1}{(\lambda_i + j - 1)!} \right| = \frac{1}{\Pi(\text{hook lengths in } [\lambda])}$$, by 19.5 and 2C

Therefore, the result will follow if we can prove:

<u>When $k \geq -h+1$, and i^* is the largest integer i such that $\lambda_i \geq k+i$, then $(d+k)^{i^*}$ divides f(d) for $k \geq 0$ and $(d+k)^{i^*+k}$ divides f(d) for $k < 0$.</u>

(k measures "how far right of the diagonal we are", and the above will ensure that the numerator in the statement of the Theorem is correct.)

<u>Case 1</u> $k \geq 0$.

For $i \leq i^*$, $d \leq d+k \leq d+\lambda_i - i$. Examining the third determinantal expression for f(d) above, we see that, for $i \leq i^*$, (d+k) divides all the entries in the ith row of our matrix. Therefore, $(d+k)^{i^*}$ divides f(d).

<u>Case 2</u> $k < 0$.

Here we claim that $f(d) = \det(M_k(d))$ where $M_k(d)$ is a matrix whos (i,j)th entry for all i, and for all $j \geq -k$, is

$$\binom{\lambda_i + d + j - i + k}{d + k} .$$

This is certainly true for $k = -1$ (by our first expression for f(d)), so assume, inductively, that it is true for k. For all $j \geq -k$, subtract the jth column of $M_k(d)$ from the (j+1)th column of $M_k(d)$. In the new matrix, for $j \geq -k+1$, the (i,j)th entry is

$$\binom{\lambda_i+d+j-i+k}{d+k} - \binom{\lambda_i+d+j-1-i+k}{d+k} = \binom{\lambda_i+d+j-i+k-1}{d+k-1}$$

Thus, our new matrix may be taken as $M_{k-1}(d)$, and the result claimed is correct.

Since $\binom{\lambda_i+j-i}{0} = \begin{cases} 0 & \text{if } \lambda_i+j-i<0 \\ 1 & \text{if } \lambda_i+j-i \geq 0 \end{cases}$

and $\lambda_i+j-i \geq 0$ for $i \leq i^*$ and $j \geq -k$, $M_k(-k)$ has the form

$$M_k(-k) = \left[\begin{array}{c|c} * & \text{1's} \\ \hline & \text{0's and 1's} \end{array}\right] \begin{array}{l} \}\, i^* \\ \}\, h-i^* \end{array}$$

$$\underbrace{\qquad}_{-k-1} \quad \underbrace{\qquad\qquad}_{h+k+1}$$

Therefore, the rank of $M_k(-k)$ is at most $(-k-1) + (h-i^*+1)$, whence the nullity of $M_k(-k)$ is at least i^*+k. Thus $(d+k)^{i^*+k}$ divides $\det(M_k(d)) = f(d)$, as required.

26.20 EXAMPLES

(i) If $\lambda = (k)$ then $\dim W^\lambda = \dfrac{d(d+1)\ldots(d+k-1)}{k!}$. In particular, $\dim W^{(2)} = \dfrac{d(d+1)}{2!}$.

(ii) If $[\lambda] = \begin{matrix} X & X & X \\ X & X & \end{matrix}$, then the hook graph is $\begin{matrix} 4 & 3 & 1 \\ 2 & 1 & \end{matrix}$

Replacing the (i,j) node in $[\lambda]$ by $j-i$, we have $\begin{matrix} 0 & 1 & 2 \\ -1 & 0 & \end{matrix}$

Then the Theorem gives $\dim W^\lambda = \dfrac{d(d+1)(d+2)(d-1)d}{4.3.2.1.1.}$

As with the Hook Formula for the dimension of the Specht module S^λ, the formula of Theorem 26.19 is much more practical than the count of semistandard tableaux when calculating dimensions of Weyl modules W^λ.

APPENDIX

THE DECOMPOSITION MATRICES OF THE SYMMETRIC GROUPS $\mathfrak{S}_n$ FOR THE PRIMES 2 AND 3 WITH $n \leq 13$

We have deliberately presented these decomposition matrices without sorting the characters into blocks. This makes it easier to spot patterns which might hold in general; for example, compare the part of the decomposition matrix of $\mathfrak{S}_{13}$ corresponding to partitions having 3 parts with the decomposition matrix of $\mathfrak{S}_{10}$, and see the remark following Corollary 24.21.

The decomposition matrices of $\mathfrak{S}_n$ for the prime 2

n = 0

		(0) 1
*1	(0)	1

n = 1

		(1) 1
*1	(1)	1

n = 2

		(2) 1
1	(2)	1
1	(1^2)	1

n = 3

		(3) 1	(2,1) 2
1	(3)	1	
*2	(2,1)		1
1	(1^3)	1	

n = 4

		(4) 1	(3,1) 2
1	(4)	1	
3	(3,1)	1	1
*2	(2^2)		1
3	(21^2)	1	1
1	(1^4)	1	

n = 5

		(5) 1	(4,1) 4	(3,2) 4
1	(5)	1		
4	(4,1)		1	
5	(3,2)	1		1
*6	(31^2)	2		1
5	(2^21)	1		1
4	(21^3)		1	
1	(1^5)	1		

n = 6

		(6) 1	(5,1) 4	(4,2) 4	(321) 16
1	(6)	1			
5	(5,1)	1	1		
9	(4,2)	1	1	1	
*16	(321)				1
10	(41^2)	2	1	1	
5	(3^2)	1		1	
10	(31^3)	2	1	1	
5	(2^3)	1		1	
9	(2^21^2)	1	1	1	
5	(21^4)	1	1		
1	(1^6)	1			

n = 7

		(7) 1	(6,1) 6	(5,2) 14	(4,3) 8	(421) 20
1	(7)	1				
6	(6,1)		1			
14	(5,2)			1		
14	(4,3)		1		1	
35	(421)	1		1		1
15	(51^2)	1		1		
21	(3^21)	1				1
21	(32^2)	1				1
*20	(41^3)		2		1	
35	(321^2)	1		1		1
14	(2^31)		1		1	
15	(31^4)	1		1		
14	(2^21^3)			1		
6	(21^5)		1			
1	(1^7)	1				

The decomposition matrix of $\mathfrak{S}_8$ for the prime 2

			(8) 1	(7,1) 6	(6,2) 14	(5,3) 8	(521) 64	(431) 40
1	(8)	(1^8)	1					
7	(7,1)	(21^6)	1	1				
20	(6,2)	$(2^2 1^4)$		1	1			
28	(5,3)	$(2^3 1^2)$		1	1	1		
64	(521)	(321^3)					1	
70	(431)	$(32^2 1)$	2	1	1	1		1
14	(4^2)	(2^4)		1		1		
21	(61^2)	(31^5)	1	1	1			
56	(42^2)	$(3^2 1^2)$	2		1			1
42	*$(3^2 2)$		2					1
35	(51^3)	(41^4)	1	2	1	1		
90	*(421^2)		2	2	2	1		1
	Block number:		1	1	1	1	2	1

The decomposition matrix of $\mathfrak{S}_9$ for the prime 2

			(9) 1	(8,1) 8	(7,2) 26	(6,3) 48	(5,4) 16	(621) 78	(531) 40	(432) 160
1	(9)	(1^9)	1							
8	(8,1)	(21^7)		1						
27	(7,2)	$(2^2 1^5)$	1		1					
48	(6,3)	$(2^3 1^3)$				1				
42	(5,4)	$(2^4 1)$			1		1			
105	(621)	(321^4)	1		1			1		
162	(531)	$(32^2 1^2)$	2		1		1	1	1	
168	(432)	$(3^2 21)$		1						1
28	(71^2)	(31^6)	2		1					
84	$(4^2 1)$	(32^3)	2		1		1		1	
120	(52^2)	$(3^2 1^3)$	2					1	1	
42	*(3^3)		2						1	
56	(61^3)	(41^5)		1		1				
189	(521^2)	(421^3)	3		2		1	1	1	
216	(431^2)	$(42^2 1)$		1		1				1
70	*(51^4)		2		2		1			
	Block number:		1	2	1	2	1	1	1	2

The decomposition matrix of $\mathfrak{S}_{10}$ for the prime 2

			(10) 1	(9,1) 8	(8,2) 26	(7,3) 48	(6,4) 16	(721) 160	(631) 198	(541) 128	(532) 200	(4321) 768
1	(10)	(1^{10})	1									
9	(9,1)	(21^8)	1	1								
35	(8,2)	(2^21^6)	1	1	1							
75	(7,3)	(2^31^4)	1		1	1						
90	(6,4)	(2^41^2)			1	1	1					
160	(721)	(321^5)						1				
315	(631)	(32^21^3)	1		2	1	1		1			
288	(541)	(32^31)						1		1		
450	(532)	(3^221^2)	2	1	1		1		1		1	
768	*(4321)											1
42	(5^2)	(2^5)			1		1					
36	(81^2)	(31^7)	2	1	1							
225	(62^2)	(3^21^4)	1		1				1			
252	(4^22)	(3^22^2)	2	1	1		1				1	
210	(43^2)	(3^31)	2	1							1	
84	(71^3)	(41^6)	2	1	1	1						
350	(621^2)	(421^4)	2	1	3	1	1		1			
567	(531^2)	(42^21^2)	3	1	3	1	2		1		1	
300	(4^21^2)	(42^3)	2	1	1	1	1				1	
525	(52^21)	(431^3)	3	1	2	1	1		1		1	
126	(61^4)	(51^5)	2	1	2	1	1					
448	$*(521^3)$							2		1		
	Block number:		1	1	1	1	1	2	1	2	1	3

The decomposition matrix of $\mathfrak{S}_{11}$ for the prime 2

			(11) 1	(10,1) 10	(9,2) 44	(8,3) 100	(7,4) 164	(6,5) 32	(821) 186	(731) 198	(641) 144	(632) 848	(542) 416	(5321) 1168
1	(11)	(1^{11})	1											
10	(10,1)	(21^8)		1										
44	(9,2)	$(2^2 1^7)$			1									
110	(8,3)	$(2^3 1^5)$		1		1								
165	(7,4)	$(2^4 1^3)$	1				1							
132	(6,5)	$(2^5 1)$				1		1						
231	(821)	(321^6)	1		1				1					
550	(731)	$(32^2 1^4)$	2				1		1	1				
693	(641)	$(32^3 1^2)$	1				1		1	1	1			
990	(632)	$(3^2 21^3)$		1		1		1				1		
990	(542)	$(3^2 2^2 1)$	2		1				1	1	1		1	
2310	(5321)	(4321^2)		3		2		2				1		1
45	(91^2)	(31^8)	1		1									
330	$(5^2 1)$	(32^4)							1		1			
385	(72^2)	$(3^2 1^5)$	1						1	1				
660	(53^2)	$(3^3 1^2)$	2		1					1			1	
462	$(4^2 3)$	$(3^3 2)$	2		1								1	
120	(81^3)	(41^7)		2		1								
594	(721^2)	(421^5)	2		1		1		1	1				
1232	(631^2)	$(42^2 1^3)$		2		3		2				1		
1155	(541^2)	$(42^3 1)$	3		1		1		1	1	1		1	
1100	$(62^2 1)$	(431^4)		2		2		1				1		
1320	$(4^2 21)$	(432^2)		2		1		1						1
1188	$*(43^2 1)$			2										1
825	(52^3)	$(4^2 1^3)$	3		1		1			1			1	
210	(71^4)	(51^6)	2		1		1							
924	(621^3)	(521^4)	2		1		1		2	1	1			
1540	(531^3)	$(52^2 1^2)$	4		1		1		2	2	1		1	
252	$*(61^5)$			2		2		1						
	Block number:		1	2	1	2	1	2	1	1	1	2	1	2

The decomposition matrix of $\mathfrak{S}_{12}$ for the prime 2

			(12) 1	(11,1) 10	(10,2) 44	(9,3) 100	(8,4) 164	(7,5) 32	(921) 320	(831) 570	(741) 1408	(651) 288	(732) 1046	(642) 416	(543) 1792	(6321) 5632	(5421) 2368
1	(12)	(1^{12})	1														
11	(11,1)	(21^{10})	1	1													
54	(10,2)	$(2^2 1^8)$		1	1												
154	(9,3)	$(2^3 1^6)$		1	1	1											
275	(8,4)	$(2^4 1^4)$	1	1		1	1										
297	(7,5)	$(2^5 1^2)$	1			1	1	1									
320	(921)	(321^7)							1								
891	(831)	$(32^2 1^5)$	3	1	1	1	1			1							
1408	(741)	$(32^2 1^3)$									1						
1155	(651)	$(32^4 1)$	1			1	1	1		1		1					
1925	(732)	$(2^5 1^2)$	3	1		1	1	1		1			1				
2673	(642)	$(3^2 2^2 1^2)$	3	1	1	1	1	1		1		1	1	1			
2112	(543)	$(3^3 21)$							1						1		
5632	(6321)	(4321^3)														1	
5775	(5421)	$(432^2 1)$	5	5	2	3	1	2		1		1	1	2			1
132	(6^2)	(2^6)				1		1									
55	(101^2)	(31^9)	1	1	1												
616	(82^2)	$(3^2 1^6)$	2		1					1							
1320	$(5^2 2)$	$(3^2 2^3)$	2		1					1		1		1			
1650	(63^2)	$(3^3 1^3)$	2	1	1	1		1					1	1			
462	(4^3)	(3^4)	2		1									1			
165	(91^3)	(41^8)	1	2	1	1											
945	(821^2)	(421^6)	3	2	2	1	1			1							
2376	(731^2)	$(42^2 1^4)$	4	2	1	3	2	2		1			1				
3080	(641^2)	$(42^3 1^2)$	4	2	1	3	2	2		1		1	1	1			
1485	$(5^2 1^2)$	(42^4)	3		1		1			1		1		1			
2079	$(72^2 1)$	(431^5)	3	2	1	2	1	1		1			1				
4158	$(53^2 1)$	$(43^2 1^2)$	2	5	1	2		1					1	1			1
2970	$(4^2 31)$	$(43^2 2)$	2	4	1	1								1			1
1925	(62^3)	$(4^2 1^4)$	3	2	1	2	1	1					1	1			
4455	(532^2)	$(4^2 21^2)$	3	5	1	3	1	2					1	1			1
2640	$*(4^2 2^2)$			4		2		1									1
330	(81^4)	(51^7)	2	2	1	1	1										
3696	(631^3)	$(52^2 1^3)$	6	2	2	3	2	2		2		1	1	1			
3520	(541^3)	$(52^3 1)$							1		1				1		
3564	$(62^2 1^2)$	(531^4)	6	2	2	2	2	1		2		1	1	1			
7700	$*(5321^2)$		8	6	2	4	2	3		2		1	2	2			1
462	(71^5)	(61^6)	2	2	1	2	1	1									
2100	$*(621^4)$		4	2	2	2	2	1		2		1					
1728	(721^3)	(521^5)							1		1						
	Block number:		1	1	1	1	1	1	2	1	2	1	1	1	2	3	1

The decomposition matrix of $\mathfrak{S}_{13}$ for the prime 2

			1 (13)	12 (12,1)	64 (11,2)	208 (10,3)	364 (9,4)	560 (8,5)	64 (7,6)	364 (1021)	570 (931)	1572 (841)	288 (751)	2848 (832)	2510 (742)	1728 (652)	2208 (643)	8008 (7321)	3200 (6421)	8448 (5431)
1	(13)	(1^{13})	1																	
12	(12,1)	(21^{11})		1																
65	(11,2)	(2^21^9)	1		1															
208	(10,3)	(2^31^7)				1														
429	(9,4)	(2^41^5)	1		1		1													
572	(8,5)	(2^51^3)		1				1												
429	(7,6)	(2^61)	1				1		1											
429	(1021)	(321^8)	1		1					1										
1365	(931)	(32^21^6)	3		1		1			1	1									
2574	(841)	(32^31^4)	4		1		1				1	1								
2860	(751)	(32^41^2)	2				1		1		1	1	1							
3432	(832)	(3^221^5)		2				1						1						
6006	(742)	$(3^22^21^3)$	4		1		1		1		2	1	1		1					
5148	(652)	(3^22^31)		1				1						1		1				
6435	(643)	(3^321^2)	3		1		1		1	1	1		1		1		1			
12012	(7321)	(4321^4)		3				2						1				1		
17160	(6421)	(432^21^2)		4		1		2						1		1		1	1	
15015	(5431)	(43^221)	7		3		1		1	1	1		1		1		1			1
66	(111^2)	(31^{10})	2		1															
1287	(6^21)	(32^5)	1				1		1		1		1							
936	(92^2)	(3^21^7)	2							1	1									
3575	(73^2)	(3^31^4)	3		1		1		1		1				1					
3432	(5^23)	(3^32^2)	2							1	1		1				1			
2574	(54^2)	(3^41)	2							1							1			
220	(101^3)	(41^9)		1		1														
1430	(921^2)	(421^7)	4		2		1			1	1									
4212	(831^2)	(42^21^5)		3		1		2						1						
6864	(741^2)	(42^31^3)	6		2		3		2		2	1	1		1					
5720	(651^2)	(42^41)		2				2						1		1				
3640	(82^21)	(431^6)		2		1		1						1						
8580	(5^221)	(432^3)		3		1		1						1		1			1	
11440	(63^21)	(43^21^3)		2		1												1	1	
3432	(4^31)	(43^3)		2		1													1	
4004	(72^3)	(4^21^5)	4		2		2		1		1				1					
12012	(632^2)	(4^221^3)		3		1		1										1	1	
12870	(542^2)	(4^22^21)	6		3		2		2		1		1		1					1
11583	(53^22)	(4^231^2)	5		3		1		1						1					1
8580	$*(4^232)$		4		2															1
495	(91^4)	(51^8)	3		2		1													
3003	(821^3)	(521^6)	5		2		1			1	1	1								
7800	(731^3)	(52^21^4)	8		2		3		2	1	3	1	1		1					
10296	(641^3)	(52^31^2)	8		2		3		2	1	3	1	2		1		1			
5005	(5^21^3)	(52^4)	3							1	1	1	1				1			
7371	(72^21^2)	(531^5)	7		2		2		1	1	3	1	1		1					
20592	(6321^2)	(5321^3)		6		1		3						2		1		1	1	
21450	(5421^2)	(532^21)	12		4		3		3	1	3	1	2		2		1			1
16016	$*(53^21^2)$		8		4		2		2		2		1		2					1
9009	(62^31)	(541^4)	7		2		2		1	1	2	1	1		1		1			
729	(81^5)	(61^7)		2		1		1												
4290	(721^4)	(621^5)	6		2		2		1	1	2	1	1							
9360	(631^4)	(62^21^3)		4		1		3						2		1				
924	$*(71^6)$		4		2		2		1											
Block number:			1	2	1	2	1	2	1	1	1	1	1	2	1	2	1	2	2	1

The decomposition matrices of $\mathfrak{S}_n$ for the prime 3

n = 0

		(0) 1
*1	(0)	1

n = 1

		(1) 1
*1	(1)	1

n = 2

		(2) 1	(1^2) 1
1	(2)	1	
1	(1^2)		1

n = 3

		(3) 1	(2,1) 1
1	(3)	1	
*2	(2,1)	1	1
1	(1^3)		1

n = 4

		(4) 1	(3,1) 3	(2^2) 1	(21^2) 3
1	(4)	1			
3	(3,1)		1		
*2	(2^2)	1		1	
3	(21^2)				1
1	(1^4)			1	

n = 5

		(5) 1	(4,1) 4	(3,2) 1	(31^2) 6	(2^21) 4
1	(5)	1				
4	(4,1)		1			
5	(3,2)		1	1		
*6	(31^2)				1	
5	(2^21)	1				1
4	(21^3)					1
1	(1^5)			1		

n = 6

		(6) 1	(5,1) 4	(4,2) 9	(3^2) 1	(41^2) 6	(321) 4	(2^21^2) 9
1	(6)	1						
5	(5,1)	1	1					
9	(4,2)			1				
5	(3^2)		1		1			
10	(41^2)		1			1		
*16	(321)	1	1		1	1	1	
9	(2^21^2)							1
5	(2^3)	1					1	
10	(31^3)					1	1	
5	(21^4)				1		1	
1	(1^6)				1			

n = 7

		(7) 1	(6,1) 6	(5,2) 13	(4,3) 1	(51^2) 15	(421) 20	(3^21) 6	(32^2) 15	(321^2) 13
1	(7)	1								
6	(6,1)		1							
14	(5,2)	1		1						
14	(4,3)			1	1					
15	(51^2)					1				
35	(421)	1		1	1		1			
21	(3^21)					1		1		
21	(32^2)		1						1	
35	(321^2)	1			1		1			1
*20	(41^3)						1			
14	(2^31)	1								1
15	(31^4)								1	
14	(2^21^3)				1					1
6	(21^5)							1		
1	(1^7)				1					

The decomposition matrix of $\mathfrak{S}_8$ for the prime 3

		1	7	13	28	1	21	35	7	35	21	90	13	28
		(8)	$(7,1)$	$(6,2)$	$(5,3)$	(4^2)	(61^2)	(521)	(431)	(42^2)	(3^22)	(421^2)	(3^21^2)	(32^21)
1	(8)	1												
7	$(7,1)$		1											
20	$(6,2)$		1	1										
28	$(5,3)$				1									
14	(4^2)			1		1								
21	(61^2)						1							
64	(521)	1			1			1						
70	(431)				1			1	1					
56	(42^2)		1	1		1				1				
*42	(3^22)						1				1			
*90	(421^2)											1		
56	(3^21^2)	1						1	1				1	
70	(32^21)		1							1				1
35	(51^3)							1						
14	(2^4)	1											1	
35	(41^4)									1				
64	(321^3)					1				1				1
28	(2^31^2)													1
21	(31^5)										1			
20	(2^21^4)								1				1	
7	(21^6)								1					
1	(1^8)					1								
Block number:		1	2	2	1	2	3	1	1	2	3	4	1	2

The decomposition matrix of $\mathfrak{S}_9$ for the prime 3

		(9) 1	(8,1) 7	(7,2) 27	(6,3) 41	(5,4) 1	(71^2) 21	(621) 35	(531) 162	(4^21) 7	(52^2) 35	(432) 21	(521^2) 189	(431^2) 27	(42^21) 189	(3^221) 41	(32^21^2) 162
1	(9)	1															
8	(8,1)	1	1														
27	(7,2)			1													
48	(6,3)		1		1												
42	(5,4)				1	1											
28	(71^2)		1				1										
105	(621)	1	1		1		1	1									
162	(531)								1								
84	(4^21)				1	1		1		1							
120	(52^2)	1	1		1	1		1			1						
168	(432)		1		1	1	1	1		1	1	1					
189	(521^2)												1				
216	(431^2)												1	1			
216	(42^21)			1											1		
168	(3^221)	1	1				1	1		1	1	1				1	
162	(32^21^2)																1
*42	(3^3)						1					1					
56	(61^3)						1	1									
84	(32^3)	1	1								1					1	
*70	(51^4)							1			1						
189	(421^3)														1		
120	(3^21^3)	1				1		1		1	1					1	
42	(2^41)	1														1	
56	(41^5)										1	1					
105	(321^4)					1				1	1	1				1	
48	(2^31^3)									1						1	
28	(31^6)									1		1					
27	(2^21^5)													1			
8	(21^7)					1				1							
1	(1^9)					1											
Block number:		1	1	2	1	1	1	1	3	1	1	1	4	4	2	1	5

The decomposition matrix of $\mathfrak{S}_{10}$ for the prime 3

		1	9	34	41	90	1	36	84	279	9	126	126	36	84	224	567	34	224	41	90	567	279
		(10)	(9,1)	(8,2)	(7,3)	(6,4)	(5^2)	(81^2)	(721)	(631)	(541)	(62^2)	(532)	(4^22)	(43^2)	(621^2)	(531^2)	(4^21^2)	(52^21)	(4321)	(3^22^2)	(42^21^2)	(3^221^2)
1	(10)	1																					
9	(9,1)		1																				
35	(8,2)	1		1																			
75	(7,3)			1	1																		
90	(6,4)					1																	
42	(5^2)				1		1																
36	(81^2)							1															
160	(721)	1		1	1				1														
315	(631)							1		1													
288	(541)									1	1												
225	(62^2)		1			1						1											
450	(532)							1		1	1		1										
252	(4^22)					1						1		1									
210	(43^2)				1		1		1						1								
350	(621^2)	1			1				1							1							
567	(531^2)																1						
300	(4^21^2)				1		1									1		1					
525	(52^21)	1		1	1		1									1			1				
*768	(4321)	1		1	1		1		1						1	1		1	1	1			
252	(3^22^2)							1					1								1		
567	(42^21^2)																					1	
450	(3^221^2)		1									1		1									1
84	(71^3)								1														
210	(3^31)	1							1						1					1			
300	(42^3)	1		1															1	1			
126	(61^4)											1											
*448	(521^3)															1			1				
525	(431^3)	1					1									1		1	1	1			
288	(32^31)		1																				1
42	(2^5)	1																		1			
126	(51^5)												1										
350	(421^4)						1								1				1	1			
225	(3^21^4)										1		1								1		
315	(32^21^3)													1									1
90	(2^41^2)																				1		
84	(41^6)														1								
160	(321^5)						1								1			1		1			
75	(2^31^4)																	1		1			
36	(31^7)													1									
35	(2^21^6)						1											1					
9	(21^8)										1												
1	(1^{10})							1															
Block numbers:		1	2	1	1	2	1	3	1	3	3	2	3	2	1	1	4	1	1	1	3	5	2

The decomposition matrix of $\mathfrak{S}_{11}$ for the prime 3

		1 (11)	10 (10,1)	34 (9,2)	109 (8,3)	131 (7,4)	1 (6,5)	45 (91^2)	120 (821)	320 (731)	693 (641)	10 (5^21)	210 (72^2)	252 (632)	45 (542)	210 (53^2)	120 (4^23)	594 (721^2)	791 (631^2)	34 (541^2)	714 (62^21)	714 (5321)	109 (4^221)	594 (43^21)	131 (432^2)	791 (52^21^2)	320 (4321^2)	693 (3^22^21)
1	(11)	1																										
10	(10,1)		1																									
44	(9,2)		1	1																								
110	(8,3)	1			1																							
165	(7,4)			1		1																						
132	(6,5)					1	1																					
45	(91^2)							1																				
231	(821)	2			1				1																			
550	(731)	1			1				1	1																		
693	(641)										1																	
330	(5^21)									1		1																
385	(72^2)		1	1		1							1															
990	(632)							1			1			1														
990	(542)										1			1	1													
660	(53^2)								1	1		1				1												
462	(4^23)					1	1						1				1											
594	(721^2)																	1										
1232	(631^2)	1							1	1									1									
1155	(541^2)									1		1							1	1								
1100	(62^21)		1	1		1	1						1								1							
2310	(5321)	2			1				1	1		1				1			1	1		1						
1320	(4^221)			1		1	2						1				1				1		1					
*1188	(43^21)																	1						1				
1320	(432^2)	2			1				1							1				1		1			1			
1540	(52^21^2)			1			1														1					1		
2310	(4321^2)		1	1			2						1				1				1		1			1	1	
990	(3^22^21)							1						1														1
120	(81^3)								1																			
825	(52^3)	2			1																	1						
462	(3^32)	1							1							1									1			
210	(71^4)												1															
924	(621^3)												1								1							
1540	(531^3)	1																	1	1		1						
825	(4^21^3)						2														1		1					
660	(3^31^2)		1										1				1										1	
1155	(42^31)		1	1																						1	1	
330	(32^4)		1																								1	
*252	(61^5)													1														
924	(521^4)															1						1						
1100	(431^4)	1										1				1				1		1			1			
1232	(42^21^3)						1										1									1	1	
990	(3^221^3)													1	1													1
693	(32^31^2)																											1
132	(2^51)	1																							1			
210	(51^6)															1												
594	(421^5)																							1				
385	(3^21^5)											1				1				1					1			
550	(32^21^4)						1										1						1				1	
165	(2^41^3)																			1					1			
120	(41^7)																1											
231	(321^6)						2										1						1					
110	(2^31^5)						1																1					
45	(31^8)														1													
44	(2^21^7)											1								1								
10	(21^9)											1																
1	(1^{11})						1																					
Block numbers:		1	2	2	1	2	2	3	1	1	3	1	2	3	3	1	2	4	1	1	2	1	2	4	1	2	2	3

The 3-regular part of the decomposition matrix of $\mathfrak{S}_{12}$ for the prime 3

		1	10	54	143	131	297	1	45	120	891	1013	10	210	252	2673	45	210	120	945	1431	1936	54	1728	1428	143	1728	945	3564	131	297	3564	1936	891	1431	1013	2673
1	(12)	1																																			
11	(11,1)	1	1																																		
54	(10,2)			1																																	
154	(9,3)	1	1		1																																
275	(8,4)	1			1	1																															
297	(7,5)						1																														
132	(6^2)					1		1																													
55	$(10\,1^2)$		1						1																												
320	(921)	2	1		1				1	1																											
891	(831)										1																										
1408	(741)	1			1	1				1		1																									
1155	(651)					1		1				1	1																								
616	(82^2)	2	1		1	1				1				1																							
1925	(732)	1	1		1	1			1	1		1		1	1																						
2673	(642)															1																					
1320	(5^22)											1	1		1		1																				
1650	(63^2)								1	1		1	1		1			1																			
2112	(543)					1		1		1		1	1	1	1		1	1	1																		
945	(821^2)																			1																	
2376	(731^2)																			1	1																
3080	(641^2)	1								1		1	1									1															
1485	(5^21^2)																				1		1														
2079	(72^21)			1			1																	1													
5632	(6321)	3	1		1	1		1	1	2		1	1	1	1			1				1			1												
5775	(5421)	2			1	1		2		1		1	2	1	1		1	1	1			1			1	1											
4158	(53^21)																			1	1		1				1										
2970	(4^231)						1																	1				1									
4455	(532^2)										1																		1								
*2640	(4^22^2)	2			1	1		2		1				1				1	1						1	1				1							
2970	(43^22)																			1							1				1						
3564	(62^21^2)																															1					
*7700	(5321^2)	3	1		1			3		1			1	1				1	1			1			2	1							1				
4455	(4^221^2)																															1		1			
4158	(43^21^2)			1																				1				1							1		
5775	(432^21)	2	2		1			2	1	1				1	1			1	1						1	1				1			1			1	
2673	$(3^22^21^2)$																																				1

The 3-singular part of the decomposition matrix of G_{12} for the prime 3

		1	10	54	143	131	297	1	45	120	891	1013	10	210	252	2673	45	210	120	945	1431	1936	54	1728	1428	143	1728	945	3564	131	297	3564	1936	891	1431	1013	2673
462	(4^3)					1		1						1					1																		
165	(91^3)								1	1																											
1925	(62^3)	2	1		1			1						1											1												
462	(3^4)	1								1								1												1							
330	(81^4)									1				1																							
1728	(721^3)																							1													
3696	(631^3)	2								1				1								1			1												
3520	(541^3)	1						2					1									1			1	1											
3520	(52^31)	2	1		1			1																	1								1				
2112	(3^321)	1	1						1	1				1	1			1	1											1						1	
1485	(42^4)			1																															1		
1320	(3^22^3)		1						1						1																					1	
462	(71^5)													1	1																						
*2100	(621^4)													1	1			1							1												
3564	(531^4)																												1								
1925	(4^21^4)	1						2					1					1							1	1				1							
3696	(52^21^3)							2										1	1						1								1				
5632	(4321^3)	1	1					3					1	1	1		1	1	2						1	1				1			1			1	
1650	(3^31^3)		1											1	1		1		1																	1	
3080	(42^31^2)		1					1											1														1			1	
1155	(32^41)	1	1																											1						1	
132	(2^6)	1																												1							
462	(61^6)														1			1																			
1728	(521^5)																										1										
2079	(431^5)																						1				1				1						
2376	(42^21^4)																											1							1		
1925	(3^221^4)							1					1		1		1	1	1							1				1						1	
1408	(32^31^3)							1											1							1				1						1	
297	(2^51^2)																														1						
330	(51^7)																	1	1																		
945	(421^6)																											1									
616	(3^21^6)							2					1					1	1							1				1							
891	(32^21^5)																																	1			
275	(2^41^4)							1																		1				1							
165	(41^8)																1		1																		
320	(321^7)							2					1				1		1							1											
154	(2^31^6)							1					1													1											
55	(31^9)												1				1																				
54	(2^21^8)																						1														
11	(21^{10})							1					1																								
1	(1^{12})							1																													
Block number:		1	1	2	1	1	2	1	1	1	3	1	1	1	1	4	1	1	1	5	5	1	5	2	1	1	5	2	3	1	5	6	1	6	2	1	7

The 3-regular part of the decomposition matrix of $\mathfrak{S}_{13}$ for the prime 3

		1	12	64	143	417	428	1	66	220	1299	1275	2287	12	495	792	5082	66	924	792	220	495	1065	4212	3367	64	1938	1428	10296	143	1938	1065	8568	417	7371	428	7371	8568	1299	3367	4212	10296	1275	2287	5082
1	(13)	1																																												
12	$(12,1)$		1																																											
65	$(11,2)$	1		1																																										
208	$(10,3)$	1		1	1																																									
429	$(9,4)$		1			1																																								
572	$(8,5)$	1			1		1																																							
429	$(7,6)$						1	1																																						
66	$(11,1^2)$								1																																					
429	$(10,2,1)$	2		1	1					1																																				
1365	(931)								1		1																																				
2574	(841)										1	1																																		
2860	(751)	1			1		1	1					1																																
1287	(6^21)											1		1																															
936	(92^2)		2			1									1																													
3424	(832)								1		1	1				1																												
6006	(742)		1			1									1		1																											
5148	(652)																	1	1																											
3575	(73^2)	1			1					1			1						1																										
6435	(643)														1		1	1		1																									
3432	(5^23)							1					1						1		1																							
2574	(54^2)											1		1		1						1																					
1430	(921^2)	2			1					1													1																					
4212	(831^2)																									1																				
6864	(741^2)	2			1								1											1		1																			
5720	(651^2)	1						1					1												1	1																		
3640	(8^221)	2		1	1		1																1				1																	
12012	(7321)	4		1	2		1	1		1			1						1				1		1		1	1																
17160	(6421)		1			1								1		1	1		1									1																	
8580	(5^221)	2			1			2					1						1		1				1	1		1		1															
11440	(63^21)	3			1			1		1			1						1				1		1	1		1			1													
15015	(5431)	2			1		1	3					1						1		1		1		1	1	1	1		1	1	1												
12012	(632^2)								1		1	1		1		1																				1											
12870	(542^2)										1	1		2		1						1												1	1											
11583	(53^22)																									1											1									
8580	*(4^232)	2			1		1	2																1				1	1		1	1	1				1									
7371	(72^21^2)																																						1									
20592	(6321^2)		2			1									1					1											1										1							
21450	(5421^2)		1			1												1		1											1										1	1						
16016	*(53^21^2)	3		1	1			3																1		1	1	1	2		1	1	1									1					
11583	(4^231^2)																																						1				1				
21450	(532^21)								1		1			1		1																1	1								1			
12870	(4^22^21)		2			1									1					1																			1	1			1		
15015	(43^221)	3		1	1			2		1									1				1					1	1		1	1	1					1				1				1	
17160	(432^21^2)								1					1		1						1														1								1			1

The decomposition matrix of $\mathfrak{S}_{13}$ for the prime 3 (continued)

		1	12	64	143	417	428	1	66	220	1299	1275	2287	12	495	792	5082	66	924	792	220	495	1065	4212	3367	64	1938	1428	10296	143	1938	1065	8568	417	7371	428	7371	8568	1299	3367	4212	10296	1275	2287	5082
220	$(10\,1^3)$									1																																			
3432	$(4^3 1)$						1	1																			1					1													
4004	(72^3)	2		1	1		1	1																			1	1																	
2574	(43^3)																						1								1					1									
495	(91^4)														1																														
3003	(821^3)																						1				1																		
7800	(731^3)	2																					1		1		1	1																	
10296	(641^3)																												1																
5005	$(5^2 1^3)$	1						2																	1	1		1		1															
9009	$(62^3 1)$		2			1																																1							
2574	$(3^4 1)$		1												1					1																							1		
5005	(52^4)	2		1	1			1																				1												1					
8580	(432^3)	2		1	1			2		1									1									1		1										1				1	
3432	$(3^3 2^2)$	1								1									1																									1	
792	(81^5)															1																													
4290	(721^4)																		1								1	1																	
9360	(631^4)															1																	1												
9009	(541^4)													2																			1	1											
9360	$(6^2 21^3)$																			1																		1							
20592	(5321^3)													2		1						1											1	1								1			
12012	$(4^2 21^3)$		1															1		1																		1	1				1		
11440	$(43^2 1^3)$	1		1				3											1		1						1	1		1		1								1				1	
10296	$(52^3 1^2)$																																									1			
6435	$(3^3 21^2)$								1							1						1																							1
5720	$(42^4 1)$	1		1				1																																1				1	
5148	$(3^2 2^3 1)$								1																																				1
1287	(32^5)		1																																								1		
*924	(71^6)																		1																										

The decomposition matrix of $\mathfrak{S}_{13}$ for the prime 3 (concluded)

		1	12	64	143	417	428	1	66	220	1299	1275	2287	12	495	792	5082	66	924	792	220	495	1065	4212	3367	64	1938	1428	10296	143	1938	1065	8568	417	7371	428	7371	8568	1299	3367	4212	10296	1275	2287	5082
4290	(621^5)																		1									1			1														
7371	(531^5)																																		1										
4004	(4^21^5)	1						2																		1		1		1	1					1									
7800	(52^21^4)							2																				1			1	1								1					
12012	(4321^4)	1						4											1		1					1		1		2	1	1				1				1				1	
3575	(3^31^4)							1											1		1									1														1	
6864	(42^31^3)							2																						1		1								1				1	
6006	$(3^22^21^3)$													1								1												1											1
2860	(32^41^2)	1						1																						1						1								1	
429	(2^61)	1																																		1									
792	(61^7)																			1																									
3003	(521^6)																														1	1													
3640	(431^6)							2																		1				1	1	1				1									
4212	(42^21^5)																																								1				
3432	(3^221^5)																	1		1																			1				1		
2574	(32^31^4)																																						1				1		
572	(2^51^3)							1																						1						1									
495	(51^8)																					1																							
1430	(421^7)							2													1									1		1													
936	(3^21^7)													2								1												1											
1365	(32^21^6)																	1																					1						
429	(2^41^5)													1																				1											
220	(41^9)																				1																								
429	(321^8)							2													1					1				1															
208	(2^31^7)							1																		1				1															
66	(31^{10})																	1																											
65	(2^21^9)							1																		1																			
12	(21^{11})													1																															
1	(1^{13})							1																																					
Block numbers:		1	2	1	1	2	1	1	3	1	3	3	1	3	2	3	2	2	1	2	1	3	1	4	1	1	1	1	2	1	1	1	3	3	4	1	5	2	2	1	5	3	2	1	3

References

1. R.W. CARTER and G. LUSZTIG, On the modular representations of the general linear and symmetric groups, Math Z. 136 (1974), 193-242.

2. C.W. CURTIS and I. REINER, "Representation theory of finite groups and associative algebras," Interscience Publishers, New York, 1962.

3. G.H. HARDY and E.M. WRIGHT, "An introduction to the theory of numbers," Oxford Univ. Press, Oxford, 1960.

4. J.S. FRAME, G. de B. ROBINSON and R.M. THRALL, The hook graphs of the symmetric group, Canad. J. Math. 6 (1954), 316-324.

5. H. GARNIR, Théorie de la representation lineaire des groupes symétriques, Mémoires de la Soc. Royale des Sc. de Liège, (4), 10 (1950).

6. G.D. JAMES, Representations of the symmetric groups over the field of order 2, J. Algebra 38 (1976), 280-308.

7. G.D. JAMES, The irreducible representations of the symmetric groups, Bull. London Math. Soc. 8 (1976), 229-232.

8. G.D. JAMES, On the decomposition matrices of the symmetric groups I, J. Algebra 43 (1976), 42-44.

9. G.D. JAMES, On the decomposition matrices of the symmetric groups II, J. Algebra 43 (1976), 45-54.

10. G.D. JAMES, A characteristic-free approach to the representation theory of $\mathfrak{S}_n$, J. Algebra 46 (1977) 430-450.

11. G.D. JAMES, On a conjecture of Carter concerning irreducible Specht modules, Math. Proc. Camb. Phil. Soc. 83 (1978), 11-17.

12. G.D. JAMES, A note on the decomposition matrices of $\mathfrak{S}_{12}$ and $\mathfrak{S}_{13}$ for the prime 3, J. Algebra, to appear.

13. A. KERBER, "Representations of permutation groups I," Lecture Notes in Mathematics, no. 240, Springer-Verlag.

14. A. KERBER and M.H. PEEL, On the decomposition numbers of symmetric and alternating groups, Mitt. Math. Sem. Univ. Giessen 91 (1971), 45-81.

15. E. MAC AOGÁIN, Decomposition matrices of symmetric and alternating groups, Trinity College Dublin Research Notes, TCD 1976-10.

16. J. McCONNELL, Note on multiplication theorems for Schur functions "Combinatoire et representation du groupe symétrique, Strasbourg 1976,"

Proceedings 1976, Ed. by D. Foata, Lecture Notes in Mathematics, no. 579, Springer-Verlag, 252-257.

17. N. MEIER and J. TAPPE, Ein neuer Beweis der Nakayama-Vermutung über die Blockstruktur Symmetrischer Gruppen, Bull. London Math. Soc. 8 (1976), 34-37.

18. M.H. PEEL, Hook representations of symmetric groups, Glasgow Math. J. 12 (1971), 136-149.

19. M.H. PEEL, Specht modules and the symmetric groups, J. Algebra 36 (1975), 88-97.

20. M.H. PEEL, Modular representations of the symmetric groups, Univ. of Calgary Research Paper no. 292, 1975.

21. D. STOCKHOFE, Die Zerlegungsmatrizen der Symmetrischen Gruppen S_{12} und S_{13} zur primzahl 2, Communications in Algebra, to appear.

22. W. SPECHT, Die irreduziblen Darstellungen der Symmetrischen Gruppe, Math Z. 39 (1935), 696-711.

23. R.M. THRALL, Young's semi-normal representation of the symmetric group, Duke J. Math. 8 (1941), 611-624.

Index